Study Guide

TO ACCOMPANY

THE HUMAN MOSAIC

Tenth Edition

Michael A. Kukral, Ph.D.
Rose-Hulman Institute of Technology

W. H. Freeman and Company
New York

ISBN-13: 978-0-7167-7256-9
ISBN-10: 0-7167-7256-6

© 2006 by W. H. Freeman and Company

All rights reserved.

Printed in the United States of America

Second printing

W. H. Freeman and Company
41 Madison Avenue
New York, NY 10010
Houndmills, Basingstoke RG21 6XS, England

CONTENTS

ACKNOWLEDGMENTS

Learning, teaching, experience, and education have always been very important in my family. This brief work is dedicated to the great teachers in my life: my parents. To the memory of my late father, Clarence Ferdinand Kukral, an upholsterer in Cleveland for 62 years, and to the memory of my mother, Ada Mae Kukral, a wonderful person from a farm in Bath, Ohio, who loved art, gardening, and people.

ABOUT THE AUTHOR

Dr. Michael A. Kukral is an Associate Professor of Geography at Rose-Hulman Institute of Technology. He received his Ph.D. from the University of Kentucky and has been distinguished for teaching excellence at both Ohio University and the University of Kentucky. Dr. Kukral was a Fulbright Scholar in the Czech capital of Prague during the "Velvet Revolution" and is the author of the book *Prague 1989: Theater of Revolution, A Study in Humanistic Political Geography,* published by Columbia University Press. He has traveled extensively in continental Europe and has taught field study courses there, as well as in Wrightsville Beach, North Carolina. Since 1999 Professor Kukral has been director of the Geography program at Rose-Hulman Institute of Technology, one of our nation's top engineering and science colleges. A native of the Cuyahoga Valley, Dr. Kukral is also a noted authority on the history, restoration, and music of the player piano and since 2003 serves as editor of the *Bulletin of the Automatic Musical Instruments Collectors' Association.* He enjoys hiking, camping, cooking, music, backpacking, and his family and grey cats Kittie and Plagle.

PREFACE

A Note to My Colleagues

There is no standard format for writing study guides. After perusing about a dozen guides and talking with many of my former and current students at the University of Kentucky, Ohio Wesleyan University, Rose-Hulman Institute of Technology, and Ohio University, I recognized two common problems that I seek to alleviate in this work.

First, many study guides replicate and summarize the textbook and are simply too lengthy (and thus an expensive addition to the textbook). Several students told me that when study guides are overly thorough, they find little reason to use or purchase the textbook. This can lead to serious problems in student comprehension of text and course material, especially given the highly visual nature of most texts in geography.

Secondly, some study guides designed for college introductory courses contain too many outside projects and assignments. The problem with this extraneous material is that it is rarely used by students. Most students in an introductory geography course are not geography majors (unfortunately) and already have enough trouble completing the assignments from their instructor. I have never met a student who has completed a study guide project on his or her own initiative.

My intention with this guide is to provide a simple and hopefully useful method for students to better understand the maps, concepts, and vocabulary of the textbook. Most students I questioned concluded that the most useful portions of any study guide were sections on key terms and the practice tests. I have expanded both of these sections from the length found in the average study guide. In addition, a new and innovative section on reading and interpreting selected maps from the textbook is included.

In summary, the value of any study guide is measured by its ability to help students better understand the textbook. Study guides cannot be written to cover other course material because non-text material is always included at the instructor's discretion. In the field of geography, as we all know, this material can be extremely diverse.

I personally welcome your comments, ideas, and especially criticisms of this study guide for the purpose of improving further editions. Please contact me directly at Kukral@Rose-Hulman.edu.

To the Student: How to Use This Guide

There is no standard method of studying that works for everyone, but there are many methods of studying that work for many people. Some students study best either while listening to music, at the library, sitting outside, with a study partner, or in the privacy of their room. Despite the location you select, it is your study methods and techniques that will determine your ability to succeed in college. Using this *Study Guide* presents one such method of textbook comprehension and retention of text information.

Follow this plan of attack for using the *Study Guide to Accompany the Human Mosaic:*

1. **Peruse** *The Human Mosaic* before beginning your first reading assignment. This basically involves looking through the book, reading some brief sections that catch your attention, and becoming familiar with the organization of the book. You should note the sections of the book such as the Contents, Preface, Index, and most importantly, the Glossary. Be sure to use the Glossary frequently!

2. When you are given a reading assignment from *The Human Mosaic,* follow this logical pattern:

    a. Peruse the chapter to get an idea what it is about.

    b. Attentively read the chapter, perhaps in sections.

    c. Take notes, ask yourself questions, and perhaps highlight key statements while reading or re-reading the chapter. (Do not highlight everything that seems important or it all becomes ugly and meaningless!)

    d. Do NOT gloss over the maps, diagrams, and photographs. These are very important to geographers and they really summarize an enormous amount of information if you interpret them correctly, thoughtfully, and thoroughly.

    e. Now you may begin using the Study Guide to enhance your textbook readings. Complete all initial sections of the Study Guide chapter **before** taking the Practice Tests.

    f. Review the textbook chapter. Now you are ready to determine your progress by taking the Practice Tests. Do NOT look for the answers in the textbook while you are taking the tests. This will not help you (it is like cheating at solitaire).

    g. Check your Practice Test answers with the answer key in the back of the *Study Guide.* Review your answer selections, especially the ones you missed, and learn from your mistakes. Rewrite your incorrect answers. Review the textbook chapter once more, rewrite your notes, get some sleep, and never miss class.

CHAPTER 1

CULTURAL GEOGRAPHY: SCIENCE AND ART

Extended Chapter Outline (including Key Terms)

I. Definition and historic development of **geography**
   A. The growth and development of geography

II. What is **cultural** geography?
   A. Learned collective behavior
   B. Spatial variation and spatial pattern
   C. Physical environments
   D. The example of wheat cultivation

III. Culture **region**
   A. Definition of region and **culture** region
   B. Maps: The essential tool
   C. **Formal** culture regions
      1. Core/periphery pattern
   D. **Functional** culture regions
      1. Nodes
   E. **Vernacular** culture regions
      1. A perceived region

IV. Cultural **diffusion**
   A. Definition of spatial spread of learned ideas
   B. Independent invention
   C. Expansion diffusion and relocation diffusion
      1. Stimulus, hierarchical, and contagious diffusion
   D. Time-distance decay
      1. Absorbing and permeable barriers
   E. Neighborhood effect

V. Cultural **ecology**
   A. Definition of cause-and-effect interplay
   B. Environmental determinism
   C. Possibilism
      1. Cultural adaptation
      2. Adaptive strategy
   D. Environmental perception
      1. The example of geomancy
      2. Natural hazards and hazard zones
   E. Humans as modifiers of the earth

VI. Cultural **interaction**
   A. Definition of systemically and spatially intertwined

1

LEARNING OBJECTIVES OF CHAPTER 1

After reading this chapter *and* studying the maps and illustrations, you should be able to:

1. Briefly trace the historical development of geography.

2. Define "geography" and how geographers approach the study of the Earth and its people.

3. Grasp the meaning of cultural geography and understand its basic topics, themes, and areas of study.

4. Describe, outline, compare, and give examples from the five major themes of cultural geography: culture region, cultural diffusion, cultural ecology, cultural integration, and cultural landscape.

5. Summarize the life and contributions of several important geographers, past and present.

6. Begin to use maps as essential tools of your education and understanding of the world.

SELECTED MAP READING AND INTERPRETATION

This section of the Study Guide is intended to heighten your map reading and interpretation skills. It will also help you apply the text readings to visual and spatial display of concepts, themes, and examples in cultural geography.

A world atlas will be very useful in completing this section of the Study Guide and will enhance your comprehension of the maps in the textbook. Ask your instructor to recommend an appropriate atlas to purchase (or visit the map collection at your library). A world atlas is essential for your personal reference library, not only during this course, but throughout your college career.

After reading the text and then studying the accompanying map and its captions, answer the following questions.

FIGURE 1.2 Areas of wheat production in the world today

1. What is the specific single trait on which this map is based?

   _____

   _____

2. What is the scale of this map?

   a. Local   b. Regional   c. Global   d. Continental

3. Are there any spatial patterns of wheat production evident at this scale? For example, is wheat cultivation stronger in the Tropics (between the Tropic of Cancer and the Tropic of Capricorn) or in the temperate latitudes (between the latitudes of 23½° and 66½° North or South)?

   _____

   _____

   _____

4. According to the map, where are the areas of significant wheat production in Africa, North America, and Russia?

   _____

   _____

   _____

FIGURE 1.4 Two-trait formal culture regions of Europe

1. What are the three major branches of Christianity portrayed on this map?

   _____, _____, _____

2. What are the three major language families on this map?

   _____, _____, _____

3. Explain what is shown and indicated by the yellow areas on this map.

_____

_____

_____

4. State the two traits for the following countries.

Spain: _____ and _____
Poland: _____ and _____
Norway: _____ and _____
Austria: _____ and _____

5. What theme or themes in cultural geography does this map illustrate?

_____

_____

_____

FIGURE 1.7 East vs. west and north vs. south in Germany

1. Which two lines of cultural-historical demarcation roughly correspond to the former Iron Curtain?

_____ and _____

2. Catholic majorities are located in Austria, _____, and _____.

3. What theme or themes in cultural geography does this map illustrate?

_____

_____

_____

FIGURE 1.8 Dixie: a vernacular region

1. How does the vernacular region of "Dixie" correspond to the "Lower South" region in Figure 1.6?

_____

_____

_____

2. Can you think of any cultural reasons why the use of the term "Dixie" is *not* as common in Virginia and Florida as in the "heart of Dixie"? Does this mean that Virginia and Florida share similar culture traits?

_____

_____

_____

3. Explain how the information was collected to create this map.

_____

_____

_____

4. Is this map a good example of core/periphery patterns or a better example of a region with nodes? Explain your reasoning.

_____

_____

_____

FIGURE 1.10 The spread of AIDS in Ohio, 1982–1990

1. Explain the hierarchical diffusion of AIDS in Ohio as shown on this map.

_____

_____

_____

2. Are the effects of contagious diffusion evident here? Describe and explain.

_____

_____

_____

3. What factors may explain the lower number of AIDS cases in southeastern Ohio?

_____

_____

_____

4. Does this map tell us that AIDS is an urban based disease? Why or why not?

_____

_____

_____

FIGURE 1.12 Chongqing and San Francisco

1. Explain the contrasting appearance of street patterns on both maps.

_____

_____

_____

2. Are either of these street patterns a good case for environmental determinism? Why or why not?

_____

_____

_____

3. Which street system is more practical? Explain.

_____

_____

_____

CREATE YOUR PERSONAL GLOSSARY OF KEY TERMS, PEOPLE, AND PLACES

In the space below, write a definition and provide an example of each key term that is sufficient for **your understanding.** It is an excellent study habit to organize your response in three parts:

1. A formal definition or identification from the textbook

2. A definition of the key term in **your own words**

3. An example to add greater understanding of the key term

KEY TERMS, PEOPLE, AND PLACES FROM CHAPTER 1:

1. geography

_____
_____
_____

2. spatial patterns

_____
_____
_____

3. analytical geographic research

_____
_____
_____

4. physical geography

_____
_____
_____

5. cultural geography

_____
_____
_____

6. culture region

_____
_____
_____

7. human traits

_____
_____
_____

8. formal culture region

_____

_____

_____

9. core/periphery pattern

_____

_____

_____

10. functional culture region

_____

_____

_____

11. cultural homogeneity

_____

_____

_____

12. political borders

_____

_____

_____

13. vernacular culture region

_____

_____

_____

14. cultural diffusion

_____

_____

_____

15. expansion diffusion

_____

_____

_____

16. relocation diffusion

_____

_____

_____

17. hierarchical diffusion

_____

_____

_____

18. contagious diffusion

_____
_____
_____

19. stimulus diffusion

_____
_____
_____

20. time-distance decay

_____
_____
_____

21. permeable barriers

_____
_____
_____

22. neighborhood effect

_____
_____
_____

23. cultural ecology

_____
_____
_____

24. environmental determinism

_____
_____
_____

25. possibilism in geography

_____
_____
_____

26. cultural adaptation

_____
_____
_____

27. environmental perception

_____
_____
_____

28. geomancy

_____

_____

_____

29. natural hazards

_____

_____

_____

30. human-as-modifier theme

_____

_____

_____

31. Judeo-Christian tradition

_____

_____

_____

32. cultural integration

_____

_____

_____

33. cultural determinism

_____

_____

_____

34. economic determinism

_____

_____

_____

35. model building

_____

_____

_____

36. culture-specific model

_____

_____

_____

37. humanistic geography

_____

_____

_____

38. "place" and "space"

_____
_____
_____

39. post-modernism

_____
_____
_____

40. cultural landscape

_____
_____
_____

41. settlement forms

_____
_____
_____

42. land-division patterns

_____
_____
_____

43. traditional American ranch-style house

_____
_____
_____

44. Yi-fu Tuan

_____
_____
_____

CHAPTER 1 REVIEW: Self-evaluation Tests

PART ONE: Multiple-choice

Circle the best answer for each question. When you are finished, read each question again with your selected answer. After you are satisfied with your practice test, use the Answer Key in the back of the Study Guide to check your responses.

1. Seven cultural geographical ideas that changed the world include maps, spatial organization, central-place theory, megalopolis, human adaptation, human transformation of the Earth, and
   a. observation.
   b. river formation.
   c. travel and exploration.
   d. sense of place.
   e. a compass.

2. The word "geography" is a Greek word meaning literally
   a. to study the Earth.
   b. to describe the Earth.
   c. to study various countries.
   d. to write about the natural environment.
   e. to analyze cultures of the world.

3. During Europe's Dark Ages, the field of academic geography was taken over by the
   a. Germans.
   b. Chinese.
   c. English.
   d. Arabs.
   e. French.

4. In the 1700s, German scholars came to view geography as the study of
   a. interrelated spatial patterns.
   b. exploration and travel.
   c. natural phenomena.
   d. landscape interpretation.
   e. ecosystems.

5. The authors of your textbook define culture as
   a. spatial variation of humans.
   b. inherited traits of humans.
   c. traditional lifestyles.
   d. learned human behavior.
   e. human perception and actions.

6. Cultural geography is the study of the spatial functioning of society and
   a. spatial variations among cultural groups.
   b. the analysis of the physical environment.
   c. human perception.
   d. the built and natural landscape.
   e. various behavioral traits of humans.

7. A culture region is a geographical unit based in
   a. vernacular characteristics.
   b. traditional patterns.
   c. human traits.
   d. complex behavioral patterns.
   e. physical properties.

8. A French-language culture region is an example of a
   a. functional culture region.
   b. formal culture region.
   c. vernacular culture region.
   d. multiple-trait area.
   e. None of the above

9. Formal regions in geography often display a core/periphery pattern.
   a. True
   b. False

10. Cultural homogeneity is the hallmark of a
    a. functional culture region.
    b. formal culture region.
    c. vernacular culture region.
    d. multiple-trait area.
    e. None of the above

11. Nodes, or central points where activities are coordinated and directed, are a common characteristic of a
    a. functional culture region.
    b. formal culture region.
    c. vernacular culture region.
    d. multiple-trait area.
    e. None of the above

12. Clearly defined borders are a common feature of a
    a. functional culture region.
    b. formal culture region.
    c. vernacular culture region.
    d. multiple-trait area.
    e. None of the above

13. Which of the following is NOT an example of a functional culture region?
    a. Coal fields of northern Appalachia
    b. Newspaper circulation area
    c. Korean-language area
    d. Cleveland metropolitan area
    e. The state of Burma

14. A vernacular culture region is often considered to be a region
    a. clearly demarcated on a map.
    b. having strong functional and formal features.
    c. perceived to exist by its inhabitants.
    d. with traditional structural traits.
    e. None of the above

15. Torsten Hagerstrand is usually associated with his ideas concerning
    a. vernacular landscape.
    b. diffusion.
    c. ethnic geography.
    d. landscape interpretation.
    e. mental maps and natural hazards.

16. The wavelike spread of ideas primarily involves the concept of
    a. stimulus diffusion.
    b. hierarchical diffusion.
    c. contagious diffusion.
    d. absorbing barriers.
    e. None of the above

17. The study of the interaction between culture and environment is termed
    a. cultural ecology.
    b. neighborhood effect.
    c. environmental science.
    d. humanistic geography.
    e. biogeography.

18. Environmental determinism was a common doctrine of belief among geographers during the
    a. mid-nineteenth century.
    b. 1950s and 1960s.
    c. late eighteenth century.
    d. early twentieth century.
    e. 1970s.

19. The belief that cultural heritage is at least as important as the physical environment in affecting human behavior is usually associated with
    a. environmental determinism.
    b. possibilism.
    c. two-trait regions.
    d. single-trait regions.
    e. material landscape.

20. Geomancy is
   a. a traditional system of land-use planning.
   b. geography based on mystical symbols.
   c. a method of weather forecasting.
   d. based primarily in Latin America.
   e. None of the above

21. The theme of cultural integration, if improperly used, can lead the geographer to
   a. social science.
   b. pre-adaptive behavior.
   c. cultural determinism.
   d. geomancy.
   e. possibilism.

22. Model building, theory, and space are terms and devices usually associated with
   the _____ approach in cultural geography.
   a. humanistic
   b. man as modifier of the Earth
   c. cultural ecological
   d. landscape interpretation
   e. social science

23. Place, topophilia, sense of place, and subjectivity are terms and ideas usually associated with
   the _____ approach in cultural geography.
   a. humanistic
   b. man as modifier of the Earth
   c. cultural ecological
   d. landscape interpretation
   e. social science

24. Most geographical studies of the cultural landscape have focused on three principal aspects:
   land-division patterns, architecture, and _____ .
   a. barn types
   b. settlement forms
   c. ethnicity
   d. food and drink
   e. spatial-temporal patterns of transportation

PART TWO: Short answer and fill-in-the-blank (probable essay-type questions)

25. The authors of the textbook are _____, _____, and
   _____.

26. The practical aspects of geography first arose among the ancient Greeks, Romans,
   _____ and _____.

15

27. Cultural geography is the study of _____
_____
_____
_____

28. The five geographical themes in cultural geography are
   a. _____
   b. _____
   c. _____
   d. _____
   e. _____

29. A culture region is a _____
_____
_____
_____

30. Explain the difference between formal culture regions and functional culture regions.
_____
_____
_____
_____

31. Give an example of a core/periphery pattern.
_____
_____
_____

32. Based on the maps in this chapter, in which culture region of North America do you now live?
_____

33. Explain why the German-language region is different from the country of Germany in the perspective of cultural geography.
_____
_____
_____

34. How are vernacular culture regions different from formal and functional regions?
_____
_____
_____

35. Explain the difference between hierarchical diffusion and stimulus diffusion.
_____
_____
_____

36. What is meant by time-distance decay?

_____

_____

_____

37. Explain the difference between environmental determinism, cultural ecology, and possibilism.

_____

_____

_____

_____

38. What is humanistic geography?

_____

_____

_____

39. Examples of natural hazards are _____

_____

_____

_____

40. What role does "perception" play in studies of cultural geography?

_____

_____

_____

_____

41. Describe the theme of cultural integration.

_____

_____

_____

_____

42. Explain the difference between the geographic concepts of "space" and "place."

_____

_____

_____

_____

43. In your own words, give an example of "placelessness."

_____

_____

_____

44. What are culture-specific models?

_____

_____

_____

45. List some actual examples of the cultural landscape.

_____

_____

_____

CHAPTER 2

MANY WORLDS: GEOGRAPHIES OF CULTURAL DIFFERENCE

Extended Chapter Outline (including Key Terms)

I. Geographies of cultural difference
    A. Many cultures
    B. Classifying cultures

II. Regions of difference
    A. Material folk culture regions
    B. Placelessness?
    C. Indigenous culture regions
    D. Food and drink
    E. Popular music
    F. Vernacular culture regions

III. Diffusion and cultural difference
    A. Agricultural fairs
    B. Blowguns
    C. Diffusion in popular culture
    D. Advertising
    E. Communications barriers
    F. Diffusion of the rodeo

IV. Ecologies of difference
    A. Indigenous ecology
    B. Local knowledge
    C. Global economy
    D. Folk ecology
    E. Gendered ecology
    F. Ecology of popular culture

V. Interaction and difference
    A. Mapping personal preferences
    B. Place images

VI. Landscapes of difference
    A. Folk architecture
    B. Folk housing in Africa south of the Sahara
    C. Folk housing in North America
    D. Landscapes of popular culture
    E. Leisure landscapes
    F. Elitist landscapes
    G. The American scene

VII.   Conclusion and further sources

LEARNING OBJECTIVES OF CHAPTER 2

After reading this chapter *and* studying the maps and illustrations, you should be able to:

1. Define, identify, and describe indigenous, folk, and popular culture and their respective regions.

2. Understand and explain various examples of folk and popular culture diffusion.

3. Describe the relationship between the environment and folk and popular cultures.

4. Provide examples and explain how various cultural practices are integrated.

5. Discuss how the cultural landscape is shaped and affected by folk and popular cultures.

6. Recognize the role of folk and popular culture in your own family, region, or heritage.

SELECTED MAP READING AND INTERPRETATION

This section of the Study Guide is intended to heighten your map reading and interpretation skills. It will also help you apply the text readings to visual and spatial display of concepts, themes, and examples in cultural geography.

A world atlas will be very useful in completing this section of the Study Guide and will enhance your comprehension of the maps in the textbook. Ask your instructor to recommend an appropriate atlas to purchase (or visit the map collection at your library or online). A world atlas is essential for your personal reference library, not only during this course, but throughout your college career.

After reading the text and then studying the accompanying map and its captions, answer the following questions.

FIGURE 2.1 Folk cultural survival regions of the United States and Canada

1. Identify the folk culture of your local region. Can you think of any examples of existing or former evidence of material folk culture?

_____

_____

_____

2. From north to south, what are the folk culture regions of the Rocky Mountains?

_____

_____

_____

3. What are the two locations of Acadian French folk culture? What is another name used for this culture in some locales?

_____

_____

_____

FIGURE 2.9 Three examples of the 40 "life-style clusters" in U.S. popular culture

1. How was the data collected for the creation of this map?

_____

_____

_____

2. What are the definitions of "Gray Power," "Old Yankee Rows," and "Norma-Rae-ville"?

_____

_____

_____

3. How do you see this map changing during the next 20 years?

_____

_____

_____

FIGURE 2.10 The Mayan culture region in Middle America

1. Identify Mayan culture regions. Do you see any relationship to contemporary international boundaries? How were the modern state boundaries created?

_____

_____

_____

2. Using an atlas, identify the climate and vegetation located in Mayan culture regions. Is it different from other areas on this map? Did the Mayans thrive in specific environments? Why or why not?

_____

_____

_____

3. Look at the location of Mayan cities. Can you identify any patterns of settlement or site selection for these cities? Based on their urban locations, would you say the Mayans were an interior or coastal oriented society? Explain the differences.

_____

_____

_____

FIGURE 2.11 The indigenous culture region of the Andes...

1. Identify the area of the Inca Empire. What current countries are located there today? Do you see any corresponding boundaries of the Inca Empire to contemporary international boundaries?

_____

_____

_____

2. The topography of the Incan culture regions shows high relief. Using an atlas, identify the various elevation changes where Inca culture exists. Would you describe the Inca as a mountain culture? What does these mean?

_____

_____

_____

3. Can you think of any reasons why the Incans didn't settle and develop the coastal areas? What environmental challenges exist in this coastal region of South America?

_____

_____

_____

FIGURE 2.12 Fast-food sales as a share of total restaurant sales, by state

1. What might account for the spatial variation in this aspect of popular culture?

    _____
    _____
    _____

2. Does this pattern bear any similarity to the map of traditional and folk cultures shown in Figure 2.1?

    _____
    _____
    _____

3. What does this suggest about the convergence hypothesis, which holds that regional cultures in America are collapsing into a national culture?

    _____
    _____
    _____

FIGURE 2.15 The vernacular "Middle West" or "Midwest"

1. What might account for the differences between this map and Figure 2.14?

    _____
    _____
    _____

2. Why would some respondents (less than 10 percent) claim to live in the Midwest in states such as Florida, California, and New Hampshire?

    _____
    _____
    _____

3. Which states on this map, such as Ohio, can be considered on the Midwest periphery?

    _____
    _____
    _____

FIGURE 2.17 Former distribution of the blowgun . . .

1. Were the two widely separated areas of blowgun use the result of independent invention or cultural diffusion?

    _____
    _____
    _____

2. Compare and contrast blowgun occurrence in the Indian and Pacific Ocean countries to the distribution of the Austronesian languages on the map in Chapter 4.

_____

_____

_____

3. Can you think of any reasons for the lack of blowgun use in continental Africa, Australia, or southern South America?

_____

_____

_____

CREATE YOUR PERSONAL GLOSSARY OF KEY TERMS, PEOPLE, AND PLACES

In the space below, write a definition and provide an example of each key term that is sufficient for **your understanding.** It is an excellent study habit to organize your response in three parts:

1. A formal definition or identification from the textbook

2. A definition of the key term in **your own words**

3. An example to add greater understanding of the key term

KEY TERMS, PEOPLE, AND PLACES FROM CHAPTER 2

1. indigenous, folk, and popular culture

_____
_____
_____

2. material culture

_____
_____
_____

3. nonmaterial culture

_____
_____
_____

5. folklore and pop culture regions

_____
_____
_____

6. material folk culture region

_____
_____
_____

7. folk and popular food regions

_____
_____
_____

8. agricultural fairs

_____
_____
_____

9. agroforestry

_____
_____
_____

10. blowguns

_____
_____
_____

11. popular and folk ecology

_____
_____
_____

12. folk medicine

_____
_____
_____

13. stock-car racing

_____
_____
_____

14. Maya, Inca, Aztec

_____
_____
_____

15. "growth, progress, fad, and trend"

_____
_____
_____

16. placelessness

_____
_____
_____

17. indigenous technical knowledge

_____
_____
_____

18. popular and folk music

_____
_____
_____

19. sports and popular culture

_____
_____
_____

20. colonialism

_____
_____
_____

21. vernacular culture and regions

_____
_____
_____

22. Wilbur Zelinsky

_____
_____
_____

23. reverse hierarchical diffusion

_____
_____
_____

24. rodeo

_____
_____
_____

25. international diffusion

_____
_____
_____

26. geography of rock and roll

_____
_____
_____

27. the convergence hypothesis

_____
_____
_____

28. mapping personal preference

_____
_____
_____

29. place image

30. elitist landscape

31. gentleman farm

32. landscape of consumption

33. landscape of leisure

34. the "American Scene"

35. residential front yards

36. sense of place

37. fast-food regions

38. sport regions

39. the vernacular "Midwest"

_____

_____

_____

40. folk and popular architecture

_____

_____

_____

41. E. Estyn Evans

_____

_____

_____

42. "New England large"

_____

_____

_____

43. "upright and wing"

_____

_____

_____

44. "saddlebag house"

_____

_____

_____

45. "shotgun house"

_____

_____

_____

46. "Ontario farmhouse"

_____

_____

_____

47. "I-house"

_____

_____

_____

CHAPTER 2 REVIEW: Self-evaluation Tests

PART ONE: Multiple-choice

Circle the best answer for each question. When you are finished, read each question again with your selected answer. After you are satisfied with your practice test, use the Answer Key in the back of the Study Guide to check your responses.

1. The best active example of a folk culture in the United States is the
   a. Hispanic culture of the Southwest.
   b. Cuban culture in Florida.
   c. Amish culture in Ohio and Pennsylvania.
   d. African-American culture in the Lower South.
   e. family farm of the Midwest.

2. Food, tools, furniture, buildings and clothing are all considered
   a. material culture.
   b. folklore .
   c. living history.
   d. folk geography.
   e. None of the above

3. Folk dialects, religions, and worldviews can be regarded as
   a. traditional lifestyles.
   b. popular culture.
   c. cultural artifacts.
   d. nonmaterial culture.
   e. None of the above

4. Active folk cultures would most likely be found in
   a. North America.
   b. Equatorial Africa.
   c. Japan and Taiwan.
   d. Scandinavia, especially Iceland.
   e. the Middle East.

5. Perhaps the most enduring feature of folk culture is
   a. architecture.
   b. language.
   c. clothing.
   d. occupation.
   e. food.

6. Folk culture spreads by the same processes of diffusion as do other elements and types of culture, but more rapidly.
   a. True
   b. False

7. Detailed knowledge about the environment by many traditional cultures is often termed
   a. cultural convergence.
   b. agricultural adaptation.
   c. indigenous technical knowledge.
   d. knowledge of ecological awareness.
   e. deep ecology.

8. If one or more nonfunctional features of blowguns occurred in both South America and Indonesia, then it is logical to surmise that the distribution of blowguns
   a. is based on the theory of independent invention.
   b. is independent of functionality.
   c. is a result of diffusion.
   d. occurred after the year 1492.
   e. none of the above.

9. A subject of folk ecology studies, the deliberate eating of Earth is called
   a. alfisol digestion.
   b. aridisol digestion.
   c. ground-food nutrition.
   d. geophagy.
   e. "rock eating."

10. In the Mexican folk culture region along the southern border of Texas, folk medicine is still widely practiced by
    a. curanderos.
    b. shamans.
    c. faith healers.
    d. witch doctors.
    e. caballeros.

11. Moonshine corn whiskey has been associated with the folk culture of the
    _____ since the 1700s.
    a. Upper Midwest
    b. Gulf Coast
    c. Rocky Mountains
    d. Great Lakes
    e. Upland South

12. Country and Western music in the United States has its origins in the folk culture of the
    a. Upper Midwest.
    b. Gulf Coast.
    c. Rocky Mountains.
    d. Great Lakes.
    e. Upland South.

13. A spatial standardization that diminishes cultural variety and demeans the human spirit can result in
    a. place image.
    b. non-place image.
    c. placelessness.
    d. aspatial values.
    e. popular culture.

14. According to the text, the leading region or places for beer consumption in the United States is
    a. the Lower South.
    b. the Upper Midwest.
    c. California and Florida.
    d. the Northeast.
    e. the Southwest.

15. Nearly all professional sports in the United States were first organized in the
    a. 1920s.
    b. 1950s.
    c. 1930s.
    d. 1890s.
    e. nineteenth century.

16. Regions perceived to exist by their inhabitants are termed
    a. popular regions.
    b. vernacular regions.
    c. folk regions.
    d. perceived regions.
    e. psychological regions.

17. The example of Wal-Mart is used illustrate the concept of
    a. reverse hierarchical diffusion.
    b. space-time distance decay.
    c. popular marketing diffusion.
    d. placelessness.
    e. None of the above

18. Time-distance decay in popular culture diffusion is _____ than/as folk culture.
    a. stronger
    b. considerably stronger
    c. weaker
    d. considerably weaker
    e. about the same

19. Popular culture is considered to have a _____ rate of diffusion.
    a. slow
    b. average
    c. rapid
    d. lateral
    e. dynamic

20. The rodeo, as an example of popular culture diffusion, has its origins in
    a. Spanish settlement areas.
    b. primarily Texas.
    c. Montana and Oklahoma.
    d. Southern California.
    e. Both answers b & c

21. This geographer, professor emeritus at Penn State, is best known for his contributions to the study of American popular culture.
    a. Ronald Isaac
    b. Michael A. Kukral
    c. John Fraser Hart
    d. Frank E. Bernard
    e. Wilbur Zelinsky

22. Of the following European countries, which probably do not participate much in Western popular culture?
    a. Czech Republic, Lithuania, Austria
    b. Switzerland, France, Luxembourg
    c. Norway, Finland, Estonia
    d. Germany, Poland, Belgium
    e. None of the above

23. Popular culture, by definition, does not display regionalization.
    a. True
    b. False

24. The devices of diffusion in popular culture require large amounts of electricity and fossil fuels.
    a. True
    b. False

25. Popular culture makes exceedingly high demands on
    a. national parks.
    b. wilderness areas.
    c. coastal zones.
    d. national recreation areas.
    e. All of the above

26. Many social scientists assume that the results of causal forces in popular culture would be to homogenize culture. This view is called
    a. placelessness.
    b. the time-distance continuum.
    c. clustering.
    d. convergence hypothesis.
    e. None of the above

27. Photography, literature, television, and film often contribute to the creation of
    a. peripheral place zones.
    b. tourist illiteracy.
    c. place image.
    d. the modernity hypothesis.
    e. None of the above

28. Which of the following would probably be labeled as a gentleman farm?
    a. Sheep ranching for wool production
    b. Horticulture of nuts and fruits
    c. Fish farming in the Mississippi
    d. Thoroughbred horse breeding in Kentucky
    e. Organic farming

29. In stage 5 of Jakle and Mattson's model of commercial strip evolution, the
    a. residential function disappears.
    b. commercial function dominates.
    c. gas stations are introduced.
    d. single-family residence dominates.
    e. None of the above

30. The most reflective landscape of consumption in popular culture is the
    a. commercial strip.
    b. shopping mall.
    c. central business district (CBD).
    d. gentleman farm.
    e. hotel and convention complexes.

31. Lowenthal concludes that the preeminence of _____ is an important part of the "American Scene."
    a. cult of bigness
    b. ugliness
    c. big hair and TV talk shows
    d. function over form
    e. manicured front lawns

32. "Bluegrass" music has its beginnings in several states in the U.S., but derives its name from the state of
    a. Texas.
    b. Kentucky
    c. Virginia.
    d. West Virginia.
    e. Tennessee.

33. Stones are often used as the traditional building material of the farmers of
    a. northern Europe.
    b. Russia.
    c. Canada.
    d. the Mediterranean.
    e. southeast Asia.

34. The Sarawak "longhouse" is a good example of traditional
    a. stone construction.
    b. half-timbering.
    c. communal housing.
    d. tropical nomadic architecture.
    e. unit farmstead.

35. A noted expert on Irish folk geography is
    a. Fred Kniffen.
    b. Estyn Evans.
    c. Connor O'Laughlin.
    d. Hubert Wilhelm.
    e. Carl Sauer.

36. An example of an African-American folk dwelling is the
    a. shotgun house.
    b. New England large.
    c. upright and wing.
    d. traditional Georgian.
    e. saddlebag house.

37. As Yankee folk migrated westward, they developed this type of house.
    a. Shotgun house
    b. New England large
    c. Upright and wing
    d. Traditional Georgian
    e. Saddlebag house

PART TWO: Short answer (probable essay-type questions)

38. In which major world regions are active folk cultures most likely to be found?
    _____
    _____
    _____

39. Provide some examples of both material and nonmaterial culture.
    _____
    _____
    _____

40. Briefly describe the difference between indigenous and folk culture.

_____
_____
_____

41. List some of the defining elements of popular culture.

_____
_____
_____

42. Provide an example and define "placelessness."

_____
_____
_____

43. What is the role of popular music in the geography of popular culture?

_____
_____
_____

44. Discuss the regional variation of beauty pageants and rodeos in the United States.

_____
_____
_____

45. What is a vernacular culture region and how do they apply to popular culture?

_____
_____
_____

46. Discuss the role of advertising in the process of diffusion.

_____
_____
_____

47. What are some barriers to the diffusion of popular culture?

_____
_____
_____

48. What are elements of Mayan and Incan culture?

_____
_____
_____

49. What are some of the environmental influences on popular culture?

_____
_____
_____

50. State some specific examples of popular culture's impact on the environment.

_____

_____

_____

51. Explain the convergence hypothesis. Do you agree with it?

_____

_____

_____

52. What is a place image and how are place images created?

_____

_____

_____

53. Provide some characteristics of an elitist landscape.

_____

_____

_____

54. What is a landscape of consumption?

_____

_____

_____

55. Provide a definition and example of a leisure landscape.

_____

_____

_____

56. Explain the meaning of Lowenthal's "The American Scene."

_____

_____

_____

57. What are some examples of popular culture in your home area?

_____

_____

_____

58. Why do you think American pop culture has become an international phenomenon?

_____

_____

_____

59. As North Americans, what features of other countries' popular culture do we absorb?

_____

_____

_____

60. What are some of the traditional folk culture regions of the western United States?

_____

_____

_____

61. Discuss some culture elements and provide an example of a folk food region.

_____

_____

_____

62. Use folk songs to demonstrate the process of folk cultural diffusion.

_____

_____

_____

63. Briefly explain the two conflicting ideas or theories concerning the global distribution of the blowgun.

_____

_____

_____

64. What is the reason behind the cultural practice of geophagy?

_____

_____

_____

65. What are three major contributions to the folk geography of the United States attributed to the Upland South region?

_____

_____

_____

66. Describe the theme of cultural integration in regards to the exchange between popular and folk culture.

_____

_____

_____

67. Explain the development of stock-car racing in the United States in terms of folk culture.

_____

_____

_____

68. Select an area of the world outside of North America and describe the traditional building materials and their relationship to the natural environment.

_____

_____

_____

CHAPTER 3

THE GEOGRAPHY OF RELIGION: SPACES AND PLACES OF SACREDNESS

Extended Chapter Outline (including Key Terms)

I. Introduction to religion
- A. Proselytic and ethnic religions
- B. Monotheism and polytheism
- C. Syncretic and othodox religions

II. Religious culture regions
- A. Judaism
- B. Christianity
  1. Western and Eastern
  2. Catholicism, Protestantism, Orthodoxy
- C. Islam
  1. Shiite
  2. Sunni
- D. Hinduism
- E. Buddhism
- F. Animism/Shamanism
- G. Secularization

III. Religious diffusion
- A. The Semitic religious hearth
- B. The Indus-Ganga hearth
- C. Barriers and time-distance decay

IV. Religious ecology
- A. Appeasing the forces of nature
- B. Environment and monotheism
  1. Ellen Churchill Semple
- C. Ecotheology
- D. Godliness and greenness

V. Cultural interaction in religion
- A. Religion and economy
  1. Food and drink taboos
- B. Religious pilgrimage

VI. Religious landscapes
- A. Religious structures
- B. Landscapes of the dead
- C. Religious names on the land
- D. Sacred space

VII. Conclusion, the Internet, and further sources

LEARNING OBJECTIVES OF CHAPTER 3

After reading this chapter *and* studying the maps and illustrations, you should be able to:

1. Understand and describe the origin and spatial diffusion of major world religions and their subdivisions.

2. Explain why certain religions expanded over several continents, such as Islam, while other religions experienced little change in range or membership size.

3. Know and describe the difference between ethnic and proselytic religions. You should also know several examples from each type.

4. Understand the regional patterns of various branches of Christianity, especially in the United States.

5. Begin to understand the role of religion in shaping politics and the economy, especially in regards to territory.

6. Discuss the major theories explaining why three major monotheistic faiths, Islam, Judaism, and Christianity, began in the same geographic region, as well as the connections between these religions.

7. Understand the various relationships between the modification of the environment and different religions.

8. Discuss aspects and locations of animistic belief systems, including their attachment to the local natural world.

9. Interpret different burial systems and the resulting landscapes of the dead.

10. Begin to think about the role of religion in shaping the history, migration, settlement patterns, and cultural geography of your home region and perhaps your family and/or ancestors.

SELECTED MAP READING AND INTERPRETATION

This section of the Study Guide is intended to heighten your map reading and interpretation skills. It will also help you apply the text readings to visual and spatial display of concepts, themes, and examples in cultural geography.

A world atlas will be very useful in completing this section of the Study Guide and will enhance your comprehension of the maps in the textbook. Ask your instructor to recommend an appropriate atlas to purchase (or visit the map collection at your library). A world atlas is essential for your personal reference library, not only during this course, but throughout your college career.

After reading the text and then studying the accompanying map and its captions, answer the following questions.

FIGURE 3.3 The world distribution of major religions

1. Other than some African countries, where is animism evident in the world?

    _____
    _____
    _____

2. What are the primarily Protestant countries of Europe?

    _____
    _____
    _____

3. Is Hinduism limited to India? If not, where else is it found?

    _____
    _____
    _____

4. Describe the changes in major religions of Africa from north to south.

    _____
    _____
    _____

FIGURE 3.4 Distribution of religious groups in Lebanon

1. What are the patterns of religious groups in Lebanon?

    _____
    _____
    _____

2. Can you think of reasons why Lebanon serves as religious refuge area? What is the pattern of the physical terrain?

_____

_____

_____

FIGURE 3.5 Leading Christian denominations in the United States and Canada

1. Can you explain the patterns of religion in Texas and Louisiana?

_____

_____

_____

2. Why do you think that Lutheranism is most prevalent in the Upper Midwest?

_____

_____

_____

3. Can you explain why some states, such as Ohio, have no denominational majority, while others such as Utah and Mississippi have basically only one majority?

_____

_____

_____

4. Other than Catholic, Baptist, and Lutheran majority areas, what other spatial patterns are evident?

_____

_____

_____

FIGURE 3.13 Secularized areas in Europe

1. Look at German speaking countries (Austria, Germany, Switzerland, Liechtenstein). What patterns prevail in secularization? Can you identify an east-west and/or north-south pattern?

_____

_____

_____

2. Can you account for any reasons or similarities among regions showing where religion is most highly practiced?

_____

_____

_____

FIGURE 3.15 The diffusion of Christianity in Europe, first to eleventh centuries

1. Look at the patterns of diffusion by the year 300. In what way do these patterns suggest hierarchical expansion diffusion?

   _____

   _____

   _____

2. What barriers to diffusion might account for the uneven advance of Christianity by the year 1050?

   _____

   _____

   _____

3. What were the last "pagan" countries of Europe? They were not Christianized until the late fourteenth century.

   _____

   _____

   _____

FIGURE 3.23 Consumption and avoidance of pork are influenced by religion

1. Can you explain the pattern in the United States? Think of factors of economy and culture.

   _____

   _____

   _____

2. How do these global patterns correspond to the map (Figure 3.3) of major world religions?

   _____

   _____

   _____

CREATE YOUR PERSONAL GLOSSARY OF KEY TERMS, PEOPLE, AND PLACES

In the space below, write a definition and provide an example of each key term that is sufficient for **your understanding.** It is an excellent study habit to organize your response in three parts:

1. A formal definition or identification from the textbook

2. A definition of the key term in **your own words**

3. An example to add greater understanding of the key term

KEY TERMS, PEOPLE, AND PLACES FROM CHAPTER 3:

1. religion

_____
_____
_____

2. proselytic religions

_____
_____
_____

3. ethnic religions

_____
_____
_____

4. sacred space

_____
_____
_____

5. spiritual geography

_____
_____
_____

6. mystical places

_____
_____
_____

7. monotheism and polytheism

_____
_____
_____

8. Western Christianity

_____
_____
_____

9. Eastern Christianity

_____
_____
_____

10. Coptic Church

_____
_____
_____

11. Maronites

_____
_____
_____

12. Nestorians

_____
_____
_____

13. Eastern Orthodoxy

_____
_____
_____

14. Protestantism

_____
_____
_____

15. "Bible Belt"

_____
_____
_____

16. Islam

_____
_____
_____

17. The Qur'an

_____

_____

_____

18. Five Pillars of Islam

_____

_____

_____

19. Shiite and Sunni Muslims

_____

_____

_____

20. Judaism

_____

_____

_____

21. Sephardim and Ashkenazim

_____

_____

_____

22. Hinduism

_____

_____

_____

23. Jainism

_____

_____

_____

24. Sikhism

_____

_____

_____

25. Buddhism

_____

_____

_____

26. Confucianism

27. Taoism

28. Shintoism and Lamaism

29. Animism and Shamanism

30. secularization

31. Semitic religious hearth

32. ethnic religions

33. contact conversion

34. Indus-Ganga religious hearth

35. religious ecology

36. geomancy or "feng-shui"

37. Ellen Churchill Semple

38. teleological view

39. doctrine of ahimsa

40. food taboos

41. "dry" counties (USA)

42. pilgrimage

43. Lourdes, France

44. Mecca, Saudi Arabia

_____
_____
_____

45. Ganga River

_____
_____
_____

46. partition of India

_____
_____
_____

47. theocracy

_____
_____
_____

48. sacred landscapes

_____
_____
_____

49. mosques and minarets

_____
_____
_____

50. religious toponyms

_____
_____
_____

CHAPTER 3 REVIEW: Self-evaluation Tests

PART ONE: Multiple-choice

Circle the best answer for each question. When you are finished, read each question again with your selected answer. After you are satisfied with your practice test, use the Answer Key in the back of the Study Guide to check your responses.

1. Religion can be defined as
   a. cultural theory.
   b. a set of beliefs.
   c. worship of either one or many gods.
   d. a way of life.
   e. a form of organized cult.

2. Religions that seek new members are termed
   a. aggressive.
   b. growth active.
   c. proselytic.
   d. charismatic.
   e. ethnic religions.

3. Judaism and Hinduism may be termed
   a. aggressive.
   b. growth active.
   c. proselytic.
   d. charismatic.
   e. ethnic religions.

4. An example of sacred space is
   a. the Dome of the Rock.
   b. a Christian church.
   c. the Wailing Wall.
   d. All of the above
   e. None of the above

5. Islam may be termed a _____ faith.
   a. polytheistic
   b. vernacular
   c. monotheistic
   d. ethnic
   e. None of the above

6. Christian Egyptians are often members of the Eastern group of Christianity called
   a. the Coptic Church.
   b. Maronites.
   c. Eastern Orthodox.
   d. Melkites.
   e. Nestorians.

7. Most of the Christians of the highland region of Lebanon are
    a. Copts.
    b. Nestorians.
    c. Greek Orthodox.
    d. Maronites.
    e. Presbyterians.

8. The core of the Mormon realm is
    a. Nevada.
    b. Missouri.
    c. South Dakota.
    d. North Dakota.
    e. Utah.

9. Which of the following is considered a prophet of Islam?
    a. Jesus
    b. Moses
    c. Muhammad
    d. Abraham
    e. All of the above

10. The stronghold of the Shiite branch of Islam is the country of
    a. Iraq.
    b. Iran.
    c. Saudi Arabia.
    d. Kuwait.
    e. Libya.

11. The Jews who eventually settled in central and eastern Europe were known as the
    a. Ashkenazim.
    b. Orthodox.
    c. Hasidic.
    d. Sephardim.
    e. None of the above

12. The majority of the world's Jewish population lives in
    a. Europe.
    b. Israel.
    c. Russia.
    d. North America.
    e. Poland and Romania.

13. The concept of "ahimsa" is focused on the
    a. caste system.
    b. Dravidian ethnic divide.
    c. idea of nirvana.
    d. veneration of all life forms.
    e. Five Pillars of Wisdom.

51

14. Hinduism has splintered into diverse religious groups that are usually regarded as separate religions. Two major direct splinter groups are
    a. Lamaism and Sikhism.
    b. Taoism and Lamaism.
    c. Coptic and Nestorian.
    d. Sikhism and Jainism.
    e. None of the above

15. Buddhism is a religion derived from Taoism.
    a. True
    b. False

16. Lamaism prevails not only in Tibet, but also in
    a. Nepal.
    b. Mongolia.
    c. China.
    d. Thailand.
    e. All of the above

17. People who are considered _____ in their form of faith believe that rocks, rivers, and other natural features can possess spirits or souls.
    a. monotheistic
    b. theocratic
    c. animistic
    d. secularized
    e. heathen

18. The youngest religion of the Semitic religious hearth is
    a. Islam.
    b. Judaism.
    c. Hinduism.
    d. Christianity.
    e. Sikhism.

19. The use of missionaries primarily involves the concept of
    a. hierarchical diffusion.
    b. relocation diffusion.
    c. distance-decay.
    d. eminent domain.
    e. All of the above

20. When political leaders (such as kings) were converted to Christianity and their subjects later followed, the process of religious diffusion was termed
    a. hierarchical diffusion.
    b. relocation diffusion.
    c. distance-decay.
    d. eminent domain.
    e. All of the above

21. Buddhism, although strongly associated with lands and people of southeastern and eastern Asia such as China and Laos, actually began in the Indian subcontinent.
   a. True
   b. False

22. A good example of absorbing barriers in religious diffusion is the attempt to spread Christianity into China.
   a. True
   b. False

23. Chinese Buddhists originally invented the magnetic compass to serve the needs of reincarnation and the travel of the soul.
   a. True
   b. False

24. The geographer often associated with environmental determinism/influence on people and their religions is
   a. Richard Hartshorne.
   b. Yi-Fu Tuan.
   c. Pradyumna P. Karan.
   d. Ellen Semple.
   e. Fred Kniffen.

25. Many followers of Islam make religious pilgrimages to the holy cities of Mecca and
   a. Jiddah.
   b. Riyadh.
   c. Damascus.
   d. Cairo.
   e. Medina.

26. In India and elsewhere the word "ghat" refers to
   a. any structure of sacred worship.
   b. a sacred space reserved for the burying of the dead.
   c. steps that often lead down to water.
   d. pagan or heathen regions of "non-believers."
   e. a balance or harmony of human and nature.

PART TWO: Short answer (probable essay-type questions)

27. Provide examples of each and discuss the difference between proselytic and ethnic religions.
   _____
   _____
   _____

28. What are some of the general spatial patterns of religion in North America?
   _____
   _____
   _____

53

29. What are some of the characteristics of the Islamic faith?

_____

_____

_____

30. What is polytheism and what are some examples of this form of religion?

_____

_____

_____

31. Briefly discuss the origin and diffusion of Buddhism, including its fusion with native ethnic religions.

_____

_____

_____

32. What are some common characteristics of animism and where can it be found today?

_____

_____

_____

33. Briefly discuss theories regarding the origin of religions in the Semitic religious hearth.

_____

_____

_____

34. Provide examples of relocation diffusion and hierarchical diffusion of religion.

_____

_____

_____

35. What religions grew out of the Indus-Ganga religious hearth?

_____

_____

_____

36. What are some absorbing barriers in the diffusion of religion?

_____

_____

_____

37. What are the practices of "feng-shui" and how do they relate to the physical environment?

_____

_____

_____

38. What is the religious significance of Mount Shasta in California?

_____
_____
_____

39. How are environmental factors used to explain the origin of monotheistic faiths?

_____
_____
_____

40. Explain the "teleological view."

_____
_____
_____

41. How can religion influence people's perception of the environment?

_____
_____
_____

42. Explain the use of fish and wine in Christianity as examples of the relationship between religion and economy.

_____
_____
_____

43. What are the various food taboos among Hindus, Muslims, Mormons, and Jews?

_____
_____
_____

44. What is the purpose of a religious pilgrimage and what are some very important pilgrimage sites for Jews, Muslims, and Christians?

_____
_____
_____

45. What role did religion play in the 1947 partition of India?

_____
_____
_____

46. How do cemeteries preserve truly ancient cultural traits? Provide some examples.

_____
_____
_____

CHAPTER 4

SPEAKING ABOUT PLACES: THE GEOGRAPHY OF LANGUAGE

Extended Chapter Outline (including Key Terms)

I.   Linguistic culture regions
    A. Language families
        1. Indo-European
        2. Afro-Asiatic
        3. Others
    B. American English dialects in the United States

II. Linguistic diffusion
    A. Indo-European diffusion
    B. Austronesian diffusion
    C. Searching for the primordial tongue
    D. Linguistic globalization

III. Linguistic ecology
    A. Habitat and vocabulary
    B. The habitat provides refuge
    C. The habitat shapes language areas

IV. Culturo-linguistic interaction
    A. Technology and linguistic dominance
    B. The social morale model
    C. The economic development model
    D. Language and religion

V.   Linguistic landscapes
    A. Messages and toponyms
    B. Generic toponyms of the United States
    C. Toponyms and cultures of the past

VI. Conclusion, Internet, and other sources

LEARNING OBJECTIVES OF CHAPTER 4

After reading this chapter *and* studying the maps and illustrations, you should be able to:

1. Explain and understand the difference between a language family, a language, a dialect, and a linguistic accent.

2. Describe the diffusion process of several major language families throughout the world.

3. Demonstrate the various relationship between language and physical environment.

4. Discuss the influence of physical and cultural environment on languages.

5. Explain several models of cultural integration involving language.

6. Understand the role of language and religion.

7. Interpret the cultural landscape through the reading of various toponyms.

8. Grasp a better understanding of the place or region where you live by interpreting local toponyms.

SELECTED MAP READING AND INTERPRETATION

This section of the Study Guide is intended to heighten your map reading and interpretation skills. It will also help you apply the text readings to visual and spatial display of concepts, themes, and examples in cultural geography.

A world atlas will be very useful in completing this section of the Study Guide and will enhance your comprehension of the maps in the textbook. Ask your instructor to recommend an appropriate atlas to purchase (or visit the map collection at your library). A world atlas is essential for your personal reference library, not only during this course, but throughout your college career.

After reading the text and then studying the accompanying map and its captions, answer the following questions.

FIGURE 4.1 Naming place is closely related to claiming place

1. What countries territorial claims overlap on this map? How does this affect place-names?
   _____
   _____
   _____

2. State some specific examples of English, Norwegian, and Spanish words on the landscape of Antarctica and match country claims with respective languages.
   _____
   _____
   _____

FIGURE 4.2 Major linguistic formal culture regions of the world

1. What is the most widespread or spatially dominant language sub-group spoken on each continent? (It is important to know the boundary between Asia and Europe!)
   _____
   _____
   _____

2. List five countries where the Austronesian language sub-group is spoken, including an African island state.
   _____
   _____
   _____

3. Looking at this map, do you see the reason behind the name "Indo-European" family? Explain.
   _____
   _____
   _____

4. Identify the language sub-group for the following independent countries: Vietnam, South Korea, Cameroon, Eritrea, Suriname, Hungary, Iceland, and Lebanon.

_____

_____

_____

FIGURE 4.8 Origin and diffusion of four major language families in the eastern hemisphere

1. What present day countries comprise the source area or hearth of the Niger-Congo language family?

_____

_____

_____

2. Where is the original source area of modern English?

_____

_____

_____

FIGURE 4.13 The environment is linguistic refuge in the Caucasus Mountains

1. The past decade has witnessed a secessionist movement in the region of Chechnya. In which independent country is Chechnya located and what is the language family sub-group of the Chechen language?

_____

_____

_____

2. Can you determine an east-west spatial pattern among two of the language family sub-groups? What is it?

_____

_____

_____

3. Does this map show greater diversity of languages at higher or lower elevations of terrain? (You will probably require an atlas to find the answer.) Can you explain the basis for the spatial patterns?

_____

_____

_____

FIGURE 4.23 Generic place names reveal the migration of Yankee New Englanders

1. Study the map. Can you think of any reasons, historical or physical, why the areas of upstate New York and northern Maine are practically devoid of the place-name types indicated on this map?

   _____

   _____

   _____

2. There is a concentration of typical place-name characteristics of New England found in northeastern Ohio. Can you think of any reasons for this concentration and why other areas of Ohio have far fewer of these toponyms?

   _____

   _____

   _____

FIGURE 4.25 Arabic toponyms in Iberia

1. Using this map, speculate concerning the direction of the Moorish invasion and retreat, the duration of Moorish Islamic rule in different parts of Iberia, and the main centers of former Moorish power.

   _____

   _____

   _____

2. Based on the text reading and using an atlas, list some present day place-names that you consider to be of Arabic origin.

   _____

   _____

   _____

3. Look at the place-names on a map of you home state or province. Are any toponyms derived from languages other than English (perhaps from Spanish, French, or Native American languages). What do these place-names reveal to you about the history of these regions?

   _____

   _____

   _____

CREATE YOUR PERSONAL GLOSSARY OF KEY TERMS, PEOPLE, AND PLACES

In the space below, write a definition and provide an example of each key term that is sufficient for **your understanding.** It is an excellent study habit to organize your response in three parts:

1. A formal definition or identification from the textbook

2. A definition of the key term in **your own words**

3. An example to add greater understanding of the key term

KEY TERMS, PEOPLE, AND PLACES FROM CHAPTER 4

1. geography of linguistics

2. languages

3. dialects

4. pidgin language

5. lingua franca

6. isoglosses

7. linguistic culture regions

8. linguistic islands

_____
_____
_____

9. language families

_____
_____
_____

10. language subfamilies

_____
_____
_____

11. Indo-European family

_____
_____
_____

12. Afro-Asiatic family

_____
_____
_____

13. polyglot

_____
_____
_____

14. Niger-Congo family

_____
_____
_____

15. Altaic family

_____
_____
_____

16. Austronesian family

_____
_____
_____

17. Sino-Tibetan family

_____
_____
_____

18. Austro-Asiatic family

_____
_____
_____

19. minor language families

_____
_____
_____

20. "regional words"

_____
_____
_____

21. "Black English"

_____
_____
_____

22. linguistic diffusion

_____
_____
_____

23. Indo-European diffusion

_____
_____
_____

24. Austronesian diffusion

_____
_____
_____

25. Polynesian people

_____
_____
_____

26. Primordial tongues

_____
_____
_____

27. Nostratic

_____
_____
_____

28. Dene-Caucasian

_____
_____
_____

29. linguistic ecology

_____
_____
_____

30. linguistic refuge areas

_____
_____
_____

31. linguistic shatter belts

_____
_____
_____

32. culturo-linguistic integration

_____
_____
_____

33. linguistic dominance

_____
_____
_____

34. Treaty of Tordesillas

_____
_____
_____

35. social morale model

_____
_____
_____

36. bilingualism

_____
_____
_____

37. monoglots

_____
_____
_____

38. economic development model

_____
_____
_____

39. changeover model

_____
_____
_____

40. linguistic landscapes

_____
_____
_____

41. toponyms

_____
_____
_____

42. generic and specific place-names

_____
_____
_____

43. the prefix "guada"

_____
_____
_____

44. the suffix "ley" or "leigh"

_____
_____
_____

45. Maori place names

_____
_____
_____

46. "loanwords"

_____
_____
_____

47. North American dialects

_____
_____
_____

48. Caucasus Mountains

_____
_____
_____

49. retreat of the Welsh language

_____
_____
_____

50. Arabic toponyms in Iberia

_____
_____

CHAPTER 4 REVIEW: Self-evaluation Tests

PART ONE: Multiple-choice

Circle the best answer for each question. When you are finished, read each question again with your selected answer. After you are satisfied with your practice test, use the Answer Key in the back of the Study Guide to check your responses.

1. Languages can be defined as
   a. speech patterns of various ethnic groups.
   b. tongues that can be mutually understood. pg 109
   c. speech patterns of singular ethnic groups.
   d. tongues that cannot be mutually understood.
   e. a collection of recognizable and similar dialects.

2. Approximately _____ languages are spoken in the world today.
   a. 850
   b. 1200
   c. 3000
   d. 4500
   e. 6000

3. One existing language may be elevated to the status of _____, or language of communication and commerce, over a wide area where it is not the mother tongue.
   a. pidgin language
   b. pidgin dialect
   c. lingua franca
   d. official or national language
   e. None of the above

4. The borders of individual word usage or pronunciations are called
   a. linguistic culture regions.
   b. isoglosses.
   c. language dependency zones.
   d. language continuum.
   e. linguistic islands.

5. Languages in the Indo-European Language family do *not* include:
   a. Turkish.
   b. German.
   c. Farsi (Persian).
   d. English.
   e. Romanian.

68

6. Three major languages of the Semitic people are Hebrew, Amharic, and
   a. Farsi (Persian).
   b. Syrian.
   c. Turkish.
   d. Arabic.
   e. Greek.

7. Swahili, an important language on East Africa, is a member of what language family?
   a. Altaic
   b. Austronesia
   c. Nilo-Saharan
   d. Niger-Congo
   e. Khoisan

8. The earliest speakers of Indo-European apparently lived in what is now
   a. India.
   b. Turkey.
   c. Russia.
   d. Germany.
   e. the Fertile Crescent.

9. The diffusion of which language family is strongly associated with island culture and vast expanses of ocean?
   a. Austronesian
   b. Indo-European
   c. Semitic
   d. Niger-Congo
   e. None of the above

10. The Spanish language, derived from Castile, as well as the Celtic tongues, are especially rich in words describing rough terrain, such as mountains.
    a. True
    b. False

11. A good example of a linguistic refuge area is
    a. Poland.
    b. Japan.
    c. southern India.
    d. South Africa.
    e. the Caucasus region.

12. Although physical barriers such as mountain ridges can slow groups from migrating from one place to another, they infrequently serve as linguistic borders.
    a. True
    b. False

13. The Treaty of Tordesillas divided _____ between Portuguese and Spanish control.
    a. Mexico
    b. the Caribbean realm
    c. South America
    d. Brazil
    e. Argentina

14. The plight of the Welsh language in the United Kingdom illustrates which of Wither's models?
    a. Social morale
    b. Clearance
    c. Economic development
    d. Changeover
    e. All of the above

15. In Muslim lands such as parts of India, Bangladesh, and Indonesia, the language of religious ceremony is
    a. the local language.
    b. the regional lingua franca.
    c. Dravidian, Bengali, and Malay.
    d. Arabic.
    e. Sanskrit.

16. Another term for place-names is
    a. call signs.
    b. landscape symbols.
    c. toponyms.
    d. signage.
    e. None of the above

17. Many place-names consist of
    a. generic parts.
    b. symbolic parts.
    c. specific parts.
    d. Both answers a and c
    e. None of the above

18. The place-name term "center" is frequently used in what American region?
    a. New England
    b. The Deep South
    c. Upper Midwest
    d. California
    e. The Southwest

19. East of the Elbe and Saale rivers, we find archaic place-name suffixes such as "ow," "in," and "zig" (as in Berlin and Leipzig). Each of these suffixes denotes a _____ origin.
    a. German
    b. Teuton
    c. Polish
    d. Celtic
    e. Slavic

20. The remnants of descriptive Arabic place-names are commonly found in regions of
    a. Italy.
    b. Greece.
    c. France.
    d. Spain.
    e. Hungary.

21. It is a fact that language is the basis for the expression of all elements of culture.
    a. True
    b. False

22. This late centenarian was best known for his pioneer studies of the linguistic geography of the eastern United States.
    a. Alfred Weber
    b. Hans Kurath
    c. Edwin Welte
    d. Alfred Kroeber
    e. James Cobban

PART TWO: Short answer (probable essay-type questions)

23. Briefly list some of the major themes encompassed in the study of the geography of language.
    _____
    _____
    _____

24. What defines a language family and what are some examples?
    _____
    _____
    _____

25. List some of the language subfamilies of the Indo-European language family.
    _____
    _____
    _____

26. Briefly trace the origin and diffusion of the Indo-European language family.
    _____
    _____
    _____

27. What are some of the barriers to language diffusion?

_____

_____

_____

28. What are the major regional dialects of the United States? Do you agree with these designations? Why or why not?

_____

_____

_____

29. Explain the diffusion of the languages of the Polynesian peoples, such as the Hawaiians, and why this diffusion merits special attention.

_____

_____

_____

30. What are the principle language families of Africa and the Middle East?

_____

_____

_____

31. Describe some of the relationships between the environment and vocabulary.

_____

_____

_____

32. How does language guide migration? Is this phenomenon true today? Why or why not?

_____

_____

_____

33. Provide some examples of both specific and generic toponyms.

_____

_____

_____

34. What are some place names in the United States derived from Native American, French, and Spanish languages? Can you think of other toponym origins in your local region?

_____

_____

_____

35. Briefly describe the various models of culturo-linguistic interaction.

_____

_____

_____

CHAPTER 5

ETHNIC GEOGRAPHY: HOMELANDS AND ISLANDS

Extended Chapter Outline (including Key Terms)

I.   Ethnic regions
    A. Ethnic culture regions in rural North America
    B. Ethnic neighborhoods and ghettos
    C. Recent ethnic migrants

II. Cultural diffusion and ethnicity
    A. Migration and ethnicity
    B. Simplification and isolation

III. Ethnic rcology
    A. Cultural preadaptation
    B. Ethnic environmental perception

IV. Ethnic cultural integration
    A. Ethnicity and livelihood
    B. Ethnic foodways
    C. Ethnicity and globalization

V.   Ethnic landscapes
    A. Visible aspects
    B. Ethnic settlement patterns
    C. Urban ethnic landscapes

VI. Conclusion

LEARNING OBJECTIVES OF CHAPTER 5

After reading this chapter *and* studying the maps and illustrations, you should be able to:

1. Present examples and explain how various groups define "ethnicity."

2. Understand the definitions of, and relationships between, national character, national origin, "nationality," race, and ethnicity.

3. Discuss the process and characteristics of ethnic migration.

4. Discuss ethnic settlement patterns, cultural preadaptation, and ethnic survival.

5. Describe the cultural integration of ethnicity and livelihood, employment, and foodways.

6. Identify and describe elements of the ethnic landscape, including settlement patterns and urban landscapes.

7. Distinguish between an ethnic homeland and an ethnic island.

8. Understand the processes involving ethnicity and assimilation, acculturation, and re-awareness.

9. Use ethnic geography as a method of understanding the process of relocation diffusion.

10. Begin to understand the role of ethnicity in your family, ancestors, and local neighborhood or region.

SELECTED MAP READING AND INTERPRETATION

This section of the Study Guide is intended to heighten your map reading and interpretation skills. It will also help you apply the text readings to visual and spatial display of concepts, themes, and examples in cultural geography.

A world atlas will be very useful in completing this section of the Study Guide and will enhance your comprehension of the maps in the textbook. Ask your instructor to recommend an appropriate atlas to purchase (or visit the map collection at your library). A world atlas is essential for your personal reference library, not only during this course, but throughout your college career.

After reading the text and then studying the accompanying map and its captions, answer the following questions.

FIGURE 5.4 Ethnic minorities in China

1. Which of the ethnic regions are homelands and which are islands?

_____

_____

_____

2. See Figure 2.1. Why are China's ethnic groups concentrated in sparsely populated peripheries of the country?

_____

_____

_____

3. Can you predict a break-up of the People's Republic of China, similar to the collapse and partition of the Soviet Union? Based on ethnicity, what new countries would you foresee?

_____

_____

_____

FIGURE 5.6 Selected ethnic homelands in North America . . .

1. What are the viable ethnic homelands and why do you think they are considered viable?

_____

_____

_____

2. What are some of the ethnic islands in the regions of ethnic island concentrations?

_____

_____

_____

3. Because they are left blank, are states such as Kentucky without ethnicity? Why or why not?

_____
_____
_____

4. Why are there no ethnic homelands indicated for Native Americans?

_____
_____
_____

FIGURE 5.7 Ethnic and national-origin groups in North America

1. What is the specific Slavic ethnic group shown in Texas?

_____
_____
_____

2. Where are the areas of Dutch ethnicity on the map?

_____
_____
_____

3. Which state shows only English ethnicity?

_____
_____
_____

4. Is there a relationship between the African ethnic area and the terrain or physical environment?

_____
_____
_____

FIGURES 5.10, 5.11, 5.12, and 5.13: Maps of ethnicity in the United States

1. Compare these four maps. What patterns emerge? Do specific states share similar statistics?

_____
_____
_____

2. Compare locations of Latino and Asian populations. What "state" patterns emerge? Why?

_____
_____
_____

3. Name some reasons for the spatial patterns by state on these maps.

_____

_____

_____

4. How do you think urbanization patterns in the United States affect ethnic patterns on these maps? Why? Explain.

_____

_____

_____

_____

FIGURE 5.20 The cultural ecology of Czech farm settlements in Texas

1. What is the relationship between Czech settlement and prairie grasses?

_____

_____

_____

2. Does it surprise you that a large Czech settlement exists in Texas? Why do you think Czechs settled here? Is it an example of preadaptation? Is the climate and soil similar to their native homeland?

_____

_____

_____

77

CREATE YOUR PERSONAL GLOSSARY OF KEY TERMS, PEOPLE, AND PLACES

In the space below, write a definition and provide an example of each key term that is sufficient for **your understanding.** It is an excellent study habit to organize your response in three parts:

1. A formal definition or identification from the textbook

2. A definition of the key term in **your own words**

3. An example to add greater understanding of the key term

KEY TERMS, PEOPLE, AND PLACES FROM CHAPTER 5:

1. "The Czech Capital of Nebraska"

_____
_____
_____

2. ethnic group

_____
_____
_____

3. host culture

_____
_____
_____

4. acculturation

_____
_____
_____

5. assimilation

_____
_____
_____

6. ethnic geography

_____
_____
_____

7. foodways

_____
_____
_____

8. ethnic minorities

9. ethnic homelands

10. ethnic islands

11. Acadiana

12. Hispanic and Latino

13. Deseret

14. Black Belt

15. ethnic substrate

16. ethnic neighborhood

17. ghetto

18. nationality

19. chain migration

20. American letters

21. channelized

22. ethnic cleansing

23. return migration

24. cultural simplification

25. cultural adaptation

26. adaptive strategy

27. preadaptation

28. first effective settlement

29. superquadra

30. aspen belt

31. tropical savanna climate

32. cultural maladaptation

33. Sami

34. economic performance

35. tortillerias

36. Dutch Calvinists

37. Hmong people

38. ethnic landscape

_____

_____

_____

39. "ethnic flag"

_____

_____

_____

40. matrilocally

_____

_____

_____

41. ethnic turf

_____

_____

_____

42. Old World

_____

_____

_____

43. Walter Kollmorgen

_____

_____

_____

44. Mennonite

_____

_____

_____

45. ethnic cuisine

_____

_____

_____

46. Little Havana

_____

_____

_____

CHAPTER 5 REVIEW: Self-evaluation Tests

PART ONE: Multiple-choice

Circle the best answer for each question. When you are finished, read each question again with your selected answer. After you are satisfied with your practice test, use the Answer Key in the back of the Study Guide to check your responses.

1. The larger society in which an ethnic group resides is referred to as
   a. an ethnic island.
   b. an ethnic majority.
   c. a host culture.
   d. a national majority.
   e. None of the above

2. When an ethnic group adopts enough of the ways of the larger society to function, this is termed
   a. assimilation.
   b. acculturation.
   c. cultural adaption.
   d. preadaptation.
   e. melting pot.

3. A complete blending with the larger society by an ethnic group is termed
   a. assimilation.
   b. acculturation.
   c. cultural adaption.
   d. preadaptation.
   e. melting pot.

4. The main difference between ethnic islands and ethnic homelands is
   a. location.
   b. spatial distribution.
   c. level of cultural integration.
   d. settlement patterns.
   e. size.

5. "Deseret" is a term used by some for the homeland of
   a. Louisiana French.
   b. French Canadians.
   c. Mexican-Americans.
   d. Jewish-Americans.
   e. the Mormons.

6. Ethnic islands are much more numerous than homelands but as common as ethnic substrates.
   a. True
   b. False

7. A ghetto is traditionally defined as a certain urban quarter where
    a. people are a racial minority.
    b. people are forced to live.
    c. people of a minority religion live.
    d. Jews traditionally live.
    e. None of the above

8. The most numerous ethnic minorities in North American cities today are originally from
    a. Africa.
    b. India.
    c. Mexico.
    d. East Asia.
    e. former communist lands.

9. When compared to the United States, which of the following ethnic or national origin groups is poorly represented in Canada?
    a. African
    b. Hispanic
    c. German
    d. Mexican
    e. All of the above

10. According to the text and tables, the largest national origin/ethnic group in the United States is
    a. Irish.
    b. English.
    c. German.
    d. African.
    e. Italian.

11. In the country of Russia, only about 81 percent of the people are ethnically Russian.
    a. True
    b. False

12. About one-third of all Canadians claim single ancestry from this ancestry group.
    a. English
    b. French
    c. German
    d. Ukrainian
    e. Scottish

13. The decision for members of an ethnic group to migrate and the actual migration usually involve
    a. hierarchical diffusion.
    b. contagious diffusion.
    c. relocation diffusion.
    d. chain migration.
    e. All of the above

14. When ethnic immigrants introduce their culture in a new land, a profound _____ occurs.
    a. prejudice
    b. cultural simplification
    c. channelization
    d. return migration
    e. All of the above

15. The example of Finnish settlement patterns in Wisconsin represents a case of
    a. preadaptation.
    b. assimilation.
    c. adaptive strategy.
    d. first effective settlement.
    e. None of the above

16. In America, the largest rural population of this ethnic group is found in Texas, where they were drawn to the ecological niche of tallgrass prairie soils.
    a. Germans
    b. Mexicans
    c. Swedes
    d. Scotch-Irish
    e. Czechs

17. The ecology of ethnic survival is often related to isolation and
    a. language.
    b. religion.
    c. elevation.
    d. food.
    e. race.

18. Light blue is a Greek ethnic color, but in Chinese urban neighborhoods the venerated and auspicious color is
    a. black.
    b. purple.
    c. green.
    d. red.
    e. yellow.

19. Green, a traditional color of ethnic Irish Catholics, is also found throughout the world in _____ neighborhoods.
    a. Japanese
    b. Russian
    c. Muslim
    d. Indian
    e. French

20. Terminology in ethnic studies is often confusing. For example, if a native white South African migrates to the United States, does he become an "African-American"? In this and many respects, definitions of ethnic identity are often ones of perception and self-recognition.
    a. True
    b. False

PART TWO: Short answer (probable essay-type questions)

21. What are some of the cultural and ethnic features that make Wilbur, Nebraska, "The Czech Capital of Nebraska"?

    _____
    _____
    _____

22. Define the terms ethnic group and host culture.

    _____
    _____
    _____

23. Explain the difference between acculturation and assimilation.

    _____
    _____
    _____

24. What is the focus of ethnic geography studies?

    _____
    _____
    _____

25. What are the two distinct geographical types of ethnic regions? Provide an example of each.

    _____
    _____
    _____

26. Provide an example of an ethnic homeland and an ethnic island in North America.

    _____
    _____
    _____

27. Explain the difference between an ethnic neighborhood and an ethnic ghetto.

    _____
    _____
    _____

28. Why are cities in North America more ethnically diverse than any other urban centers in the world?

_____

_____

_____

29. What are the top five ethnic ancestry/national origin groups in the United States? How is this different from Canada?

_____

_____

_____

30. Explain the concept of chain migration, including an example of an "American letter."

_____

_____

_____

31. Briefly discuss the concept of preadaptation in ethnic migration.

_____

_____

_____

32. Provide some examples of ethnic environmental perception.

_____

_____

_____

33. What is the relationship between ethnicity and business activity?

_____

_____

_____

34. What is the relationship between ethnicity and type of employment?

_____

_____

_____

35. Who and where are the Hmong-Americans and what are their distinctive gardening practices?

_____

_____

_____

36. Other than the Finnish sauna, what is an example of an "ethnic flag"?

_____

_____

_____

37. How can colors connote and reveal ethnicity? Provide examples.

_____
_____
_____

38. What are some of the contributions of Professor Walter Kollmorgen?

_____
_____
_____

39. What are some local examples of ethnicity in your home town, city, or region.

_____
_____
_____

40. Nearly everyone in North America is part of some ethnic fabric or identity. Describe the role of ethnicity or ethnic ancestry in your family or acquaintances.

_____
_____
_____

CHAPTER 6

POLITICAL GEOGRAPHY: A DIVIDED WORLD

Extended Chapter Outline (including Key Terms)

I. Political culture regions
   A. Independent countries
      1. Territoriality
      2. Boundaries
      3. Centrifugal and centripetal forces
   B. Supranational political bodies
   C. Electoral geography

II. Political diffusion
   A. Country building as diffusion
   B. Diffusion of independence and innovations

III. Political ecology
   A. Geopolitics
   B. The heartland theory
      1. Halford Mackinder
   C. Warfare and environmental destruction

IV. Politico-cultural interaction
   A. The nation-state
   B. Ethnic separatism
   C. The cleavage model
   D. Example: Sakha Republic
   E. Political imprint on economic geography

V. Political landscapes
   A. Imprint on the legal code
      1. Border landscapes
   B. Physical properties of boundaries
   C. The impress of central authority
   D. National iconography on the landscape

VI. Conclusion

LEARNING OBJECTIVES OF CHAPTER 6

After reading this chapter *and* studying the maps and illustrations, you should be able to:

1. Describe various types of states, countries, and other political bodies.

2. Understand the themes and concepts associated with country building.

3. Explain the political ecology of folk fortresses, warfare and environmental destruction, and Mackinder's heartland theory.

4. Explain the relationship of ethnicity and territory to the nation-state and the multinational state.

5. Describe the effects of political decision making on the cultural landscape.

6. Provide examples of the basic elements of the political landscape such as the boundary.

7. Better understand many of the world's current political conflicts and wars based on your knowledge of political and cultural geography.

SELECTED MAP READING AND INTERPRETATION

This section of the Study Guide is intended to heighten your map reading and interpretation skills. It will also help you apply the text readings to visual and spatial display of concepts, themes, and examples in cultural geography.

A world atlas will be very useful in completing this section of the Study Guide and will enhance your comprehension of the maps in the textbook. Ask your instructor to recommend an appropriate atlas to purchase (or visit the map collection at your library). A world atlas is essential for your personal reference library, not only during this course, but throughout your college career.

After reading the text and then studying the accompanying map and its captions, answer the following questions:

FIGURE 6.1 The independent countries of the world

1. The United States basically shares a border with only two countries. How many countries now border Russia and what are they? (Don't forget Norway!)

   _____

   _____

   _____

2. What are the newly formed independent countries of Central Asia and the Caucasus region that were formerly part of the Soviet Union?

   _____

   _____

   _____

3. What independent countries were recently derived from the break-up of Yugoslavia, Ethiopia, and Czechoslovakia?

   _____

   _____

   _____

4. Locate these independent countries on the world map: Bhutan, Sierra Leone, Oman, French Guiana, Slovenia, Lithuania, Belize, Papua New Guinea, Swaziland, Haiti, and Vanuatu.

FIGURE 6.3 Two independent countries, A and B

1. These are real countries. Using an atlas, identify them.

   _____

   _____

   _____

2. The conditions shown and described here are for 1994. What was the territorial outcome of this dispute?

_____

_____

_____

3. What and where are the actual religions, languages, and ethnicity of the various territories shown on this map?

_____

_____

_____

FIGURE 6.5 *Some supranational political organizations in the Eastern Hemisphere*

1. Can you identify the reasons why certain European countries, such as Bosnia or Moldova, are *not* part of the European Union?

_____

_____

_____

2. If British colonialism has basically ended, why are certain countries such as Zimbabwe and India part of the British Commonwealth?

_____

_____

_____

3. What territories of the former Soviet Union are *not* parts of the Commonwealth of Independent States? Why not?

_____

_____

_____

FIGURE 6.8 *Russia developed from a core area*

1. Can you think of reasons why expansion to the east was greater than to the west? What were the barriers?

_____

_____

_____

2. What environmental goals might have motivated Russian expansion?

_____

_____

_____

3. What were the causes behind the recent contraction of the country? Do you foresee any further contraction?

_____

_____

_____

FIGURE 6.9 Independence from European colonial rule...

1. Can you identify some barriers that slowed the diffusion of independence in Africa?

_____

_____

_____

2. 1960 was a landmark year for African independence. Using other maps, can you identify areas of British, French, Belgian, and Portuguese rule in Africa and show if a relationship exists between colonial ruler and year of independence?

_____

_____

_____

3. Why did independence come so late to Namibia? Can you find any reasons for this? Who were the colonial rulers of Namibia?

_____

_____

_____

4. List all the countries achieving independence before 1960 and their respective year of sovereignty.

_____

_____

_____

FIGURE 6.16 Nation-states, multinational countries, and other types

1. Identify 10 nation-states from this map.

_____

_____

_____

2. Why are the countries of Canada and Spain classified as old multinational states?

_____

_____

_____

3. What is your opinion on the United States as a country evolving toward nation-statehood?

_____

_____

_____

4. This classification is arbitrary and debatable. How would you change it, and why?

_____

_____

_____

FIGURE 6.18: Kurdistan

1. In what countries are areas of Kurdish predominance found?

_____

_____

_____

2. Using your atlas, can you identify the physical landscape of "Kurdistan" and thus show that a folk fortress situation exists?

_____

_____

_____

3. Why and how does this map show reasons for political instability in this region? Is this especially true for Iraq and Iran? Why or why not?

_____

_____

_____

4. If Kurdistan becomes an independent country, what problems of a geopolitical nature, such as access to the seas, may they experience?

_____

_____

_____

CREATE YOUR PERSONAL GLOSSARY OF KEY TERMS, PEOPLE, AND PLACES

In the space below, write a definition and provide an example of each key term that is sufficient for **your understanding.** It is an excellent study habit to organize your response in three parts:

1. A formal definition or identification from the textbook

2. A definition of the key term in **your own words**

3. An example to add greater understanding of the key term

KEY TERMS, PEOPLE, AND PLACES FROM CHAPTER 4:

1. political geography

_____
_____
_____

2. independent countries

_____
_____
_____

3. political culture region

_____
_____
_____

4. territoriality

_____
_____
_____

5. nationalism

_____
_____
_____

6. national territory

_____
_____
_____

7. enclave and exclave

_____
_____
_____

8. "shoestring" countries

_____
_____
_____

9. boundaries

_____
_____
_____

10. marchlands

_____
_____
_____

11. buffer state

_____
_____
_____

12. satellite state

_____
_____
_____

13. natural boundaries

_____
_____
_____

14. ethnographic boundaries

_____
_____
_____

15. geometric boundaries

_____
_____
_____

16. unitary and federal spatial organization

_____
_____
_____

17. centripetal forces

_____
_____
_____

18. centrifugal forces

_____
_____
_____

19. raison d'etre

_____
_____
_____

20. supranational organizations

_____
_____
_____

21. electoral geography

_____
_____
_____

22. cleavages

_____
_____
_____

23. gerrymandering

_____
_____
_____

24. political diffusion

_____
_____
_____

25. secession movements

_____
_____
_____

26. diffusion of insurgencies and innovations

_____
_____
_____

27. political ecology

_____
_____
_____

28. folk fortress

_____
_____
_____

29. manifest destiny

_____
_____
_____

30. heartland theory

_____
_____
_____

31. Halford Mackinder

_____
_____
_____

32. rimland

_____
_____
_____

33. politico-cultural integration

_____
_____
_____

34. nation-state

_____
_____
_____

35. multinational country

36. ethnic separatism

37. cleavage model

38. Kurdistan

39. political imprint

40. political landscape

41. imprint of the legal code

42. border landscapes

43. impress of central authority

44. "military landscapes"

_____

_____

_____

45. national iconography

_____

_____

_____

46. territorial imperative

_____

_____

_____

47. Kurdistan

_____

_____

_____

48. Great Wall of China

_____

_____

_____

49. Mount Rushmore

_____

_____

_____

CHAPTER 6 REVIEW: Self-evaluation Tests

PART ONE: Multiple-choice

Circle the best answer for each question. When you are finished, read each question again with your selected answer. After you are satisfied with your practice test, use the Answer Key in the back of the Study Guide to check your responses.

1. The Earth's surface is divided into approximately _____ independent countries.
   a. 230
   b. 320
   c. 280
   d. 200
   e. 190

2. Europe and Africa each have about the same number of independent countries.
   a. True
   b. False

3. Theoretically, the most desirable shape for a country is
   a. elongated.
   b. square.
   c. hexagonal.
   d. triangular.
   e. None of the above

4. These areas are pieces of national territory separated from the main body of a country by the territory of another.
   a. Peninsulas
   b. Exclaves
   c. Enclaves
   d. Colonies
   e. Protectorates

5. Until quite recently, many boundaries were not clear or sharp, but undefined, somewhat fuzzy zones called
   a. buffer states.
   b. frontiers.
   c. hinterlands.
   d. marchlands.
   e. international zones.

6. Mongolia and Nepal serve as good examples of
   a. "shoestring" countries.
   b. buffer states.
   c. secessionist states.
   d. satellite states.
   e. rump or truncated states.

7. Boundaries that are based on neither cultural nor physical features are often
   a. ethnographic.
   b. geometric.
   c. natural.
   d. irregular.
   e. None of the above

8. An excellent example of a relic boundary exists
   a. between Canada and Alaska.
   b. between Brazil and Argentina.
   c. between Laos and Cambodia.
   d. within Germany.
   e. within Ireland.

9. A federal government is usually considered a less geographically expressive system.
   a. True
   b. False

10. The United States, Canada, France, and Australia are all examples of federal governments.
    a. True
    b. False

11. Whatever disrupts internal order and encourages destruction of the country is called
    a. raison d'etre.
    b. insurgency.
    c. centrifugal force.
    d. nationalism.
    e. centripetal force.

12. Electoral geography is useful for identifying
    a. the Lower South.
    b. formal and functional culture regions.
    c. spatial patterns of ethnicity.
    d. supranational organizations.
    e. None of the above

13. The creation of districts that have a majority of voters favoring the group in power and a minority of opposition voters is called
    a. gerrymandering.
    b. cleavage control districts.
    c. redistricting.
    d. raison d'etre.
    e. selective electoral districts.

14. During political diffusion, the original core area seldom remains the country's most important district.
    a. True
    b. False

15. The Russian state originated in the small principality of
    a. St. Petersburg.
    b. Kiev.
    c. Warsaw.
    d. Novgorad.
    e. Moscow.

16. Potentially, countries without political core areas, such as Zaire and Belgium, are the least stable of all independent states.
    a. True
    b. False

17. The best example of contagious expansion diffusion in political geography is
    a. the growth of the Russian Empire.
    b. political independence in Africa.
    c. the creation of Mexico.
    d. British colonization of India.
    e. the reunification of Germany.

18. Throughout much of history, a country's survival was enhanced by some sort of natural protection. These areas of protection are called
    a. national moats.
    b. natural defenses.
    c. ethnic islands.
    d. national shields.
    e. folk fortresses.

19. The "heartland theory" was developed by geographer
    a. Wilbur Zelinsky.
    b. Frank Ainsley.
    c. Halford Mackinder.
    d. Jean Gottmann.
    e. Derwent Whittlese.

20. The "heartland theory" predicted, in effect
    a. the growth of Southern Democrats.
    b. the economic power of the Midwest.
    c. the rise of French and British colonialism.
    d. Russian conquest of the world.
    e. communism in mainland China.

21. Which of the following is *not* an example of a modern nation-state?
    a. Sweden
    b. Belgium
    c. Japan
    d. Armenia
    e. Germany

22. The greatest concentration of the francophone cultural-linguistic minority in Canada is found in
    a. Toronto.
    b. Ontario.
    c. Manitoba.
    d. Newfoundland and New Brunswick.
    e. Quebec.

23. The "impress of central authority" refers to
    a. usually unitary rather than federal states.
    b. transportation networks.
    c. omnipresent military authority.
    d. government and landscape.
    e. None of the above

24. A good example of national iconography in the American landscape is
    a. bald eagles and flags.
    b. the National Cathedral.
    c. the Grand Canyon.
    d. New York City.
    e. baseball.

25. This former principality is used as an example for illustrating the relationship between terrain and political geography
    a. Andorra.
    b. Liechtenstein.
    c. Bavaria.
    d. Berchtesgaden.
    e. Denmark.

PART TWO: Short answer (probable essay-type questions)

26. Briefly describe the meaning of the "territorial imperative."

    _____
    _____
    _____

27. How and why are political boundaries barriers to cultural diffusion?

    _____
    _____
    _____

28. State some of the contributions of geographer Halford Mackinder.

    _____
    _____
    _____

29. Briefly explain the process of contagious expansion diffusion by using the example of political independence in Africa.

_____

_____

_____

30. What are folk fortresses and are they significant today?

_____

_____

_____

31. Briefly summarize the importance of the heartland theory.

_____

_____

_____

32. What are some of the environmental catastrophes associated with the Persian Gulf War?

_____

_____

_____

33. What is the difference between a nation-state and a multinational country? Use examples from each.

_____

_____

_____

34. Briefly explain the cleavage model in political geography.

_____

_____

_____

35. How is the legal code imprinted on landscapes? Use examples.

_____

_____

_____

36. Describe the various types of boundaries between and within countries.

_____

_____

_____

37. What is some of the local political iconography of your state, province, or region?

_____

_____

_____

38. Define the term "territoriality."

_____

_____

_____

39. Provide some examples and explain the difference between exclaves and enclaves.

_____

_____

_____

40. Briefly describe and explain the difference between unitary and federal spatial organization of government.

_____

_____

_____

CHAPTER 7

GEODEMOGRAPHY: PEOPLING THE EARTH

Extended Chapter Outline (including Key Terms)

I. Demographic regions
   A. Population distribution
      1. Population density
      2. Ten most populous countries
      3. Thickly, moderately, and thinly settled areas
      4. Absolute and physiological density
   B. Patterns of natality
      1. Birthrates
      2. Fertility rates
   C. The geography of mortality
      1. Death rates
   D. The population explosion
      1. Crucial elements
      2. Geometric progression
      3. Malthus
      4. Demographic stabilization
   E. Demographic transformation
   F. Age distributions
      1. Population pyramid
   G. Geography of gender
      1. Sex ratio
      2. Gendered spaces
   H. Standard of living

II. Diffusion in population geography
   A. Migration
      1. Voluntary and forced migration
      2. Push-and-pull factors
   B. Disease diffusion
      1. Example of AIDS
   C. Diffusion of fertility control

III. Population ecology
   A. Environmental influence
      1. Climatic factors
      2. Coastal locations
   B. Environmental perception and population distribution
      1. Interregional migration
   C. Population density and environmental alteration
      1. Changes in vegetation

IV. Cultural integration and population patterns
   A. Cultural factors
      1. Food, crops, attitudes
      2. Influence of gender
      3. Personal space
   B. Political factors
      1. Government actions
   C. Economic factors
   D. Gender and geodemography

V. The settlement landscape
   A. Farm villages
      1. Clustered settlements
      a. Village pattern types
      2. Environmental determinants
   B. Isolated farmsteads
      1. Global locations
   C. Reading the cultural landscape
      1. Example of Spanish and Mayan influence

VI. Conclusion

LEARNING OBJECTIVES OF CHAPTER 7

After reading this chapter *and* studying the maps and illustrations, you should be able to:

1. Understand the terms and concepts used by geographers to study the human population.

2. Give examples of the various factors influencing human migration and population diffusion.

3. Distinguish and discuss the differences among population distribution, population density, and population composition.

4. Show the relationship between population and the natural environment, including factors of human perception.

5. Understand the role of political, economic, and especially cultural phenomena in influencing the human population.

6. Distinguish various settlement patterns and farm village patterns.

7. Continue to improve your ability to read and interpret the cultural landscape.

SELECTED MAP READING AND INTERPRETATION

This section of the Study Guide is intended to heighten your map reading and interpretation skills. It will also help you apply the text readings to visual and spatial display of concepts, themes, and examples in cultural geography.

A world atlas will be very useful in completing this section of the Study Guide and will enhance your comprehension of the maps in the textbook. Ask your instructor to recommend an appropriate atlas to purchase (or visit the map collection at your library). A world atlas is essential for your personal reference library, not only during this course, but throughout your college career.

After reading the text and then studying the accompanying map and its captions, answer the following questions.

FIGURE 7.1 Population density of the world

1. It is well-known that China has the largest population in the world. The pattern of population distribution in China, however, shows a clear east-west division. Can you think of any environmental or economic factors influencing this pattern in China? Secondly, do you see a similar distribution pattern in the United States, Brazil, and Australia? Provide a few statements to support your answer.

   _____

   _____

   _____

   _____

2. Is it reasonable to state that population density is usually highest in coastal regions? Select a few specific countries or regions to support your answer.

   _____

   _____

   _____

3. What is the pattern of population density in the tropics? Use the continents of Africa, Asia, and South America in your answer.

   _____

   _____

   _____

4. Can you think of any reasons for the band of high population density running across the interior of Central Europe?

   _____

   _____

   _____

FIGURE 7.2 The total fertility rate (TFR) in the world

1. What regions of the world have the highest TFR?

    _____

    _____

    _____

2. Can you think of any reasons why some Latin American countries, such as Honduras, have higher TFR than others?

    _____

    _____

    _____

3. Compare the data on this map with the data shown in Figure 7.1. Does this comparison illustrate that the population of China and India must be declining? Why or why not?

    _____

    _____

    _____

FIGURE 7.3 The Geography of HIV/AIDS

1. What is the population density of the countries with a high rate of HIV/AIDS?

    _____

    _____

    _____

2. Where is HIV/AIDS lowest according to this map? Can you make some suggestions for the low numbers? What role does government have in reporting these numbers?

    _____

    _____

    _____

3. Can you find a relationship between HIV/AIDS and the economy of a region or country? Any examples?

    _____

    _____

    _____

FIGURE 7.8 Geography of contraception in the modern world

1. Where do the highest percentages occur? Are these rich or poor areas?

    _____

    _____

    _____

111

2. The white areas on this map are an indication of what?

_____

_____

_____

3. Compare this map with Figure 7.2. What can you interpret from this comparison?

_____

_____

_____

FIGURE 7.23 Protein malnutrition, vegetarianism, and rice consumption in India

1. Do these maps show a relationship between vegetarianism and malnutrition in India? Why or why not?

_____

_____

_____

2. Is there an evident relationship between predominantly rice consuming areas and malnutrition?

_____

_____

_____

3. Can you think of any cultural, economic, or environmental reasons for the patterns on these maps?

_____

_____

_____

CREATE YOUR PERSONAL GLOSSARY OF KEY TERMS, PEOPLE, AND PLACES

In the space below, write a definition and provide an example of each key term that is sufficient for **your understanding.** It is an excellent study habit to organize your response in three parts:

1. A formal definition or identification from the textbook

2. A definition of the key term in **your own words**

3. An example to add greater understanding of the key term

KEY TERMS, PEOPLE, AND PLACES FROM CHAPTER 7:

1. demography

_____
_____
_____

2. population density

_____
_____
_____

3. demographic regions

_____
_____
_____

4. population distribution

_____
_____
_____

5. thickly, moderately, and thinly settled areas

_____
_____
_____

6. physiological density

_____
_____
_____

7. absolute density

_____
_____
_____

8. natality

_____
_____
_____

9. birthrates

_____
. _____
_____

10. fertility rate

_____
_____
_____

11. mortality

_____
_____
_____

12. death rates

_____
_____
_____

13. population explosion

_____
_____
_____

14. Thomas Malthus

_____
_____
_____

15. demographic stabilization

_____
_____
_____

16. demographic transformation

_____
_____
_____

17. post-industrial period

_____
_____
_____

18. population pyramid

_____
_____
_____

19. sex ratio

_____
_____
_____

20. gendered spaces

_____
_____
_____

21. migration

_____
_____
_____

22. voluntary and forced migration

_____
_____
_____

23. push-and-pull factors

_____
_____
_____

24. diffusion of fertility control

_____
_____
_____

25. disease diffusion

_____
_____
_____

26. AIDS, HIV-1, HIV-2

_____
_____
_____

27. population ecology

_____
_____
_____

28. preadaptation

_____
_____
_____

29. sustainability

_____
_____
_____

30. European coal fields

_____
_____
_____

31. interregional migration

_____
_____
_____

32. kitchen gardens

_____
_____
_____

33. matrilineal societies

_____
_____
_____

34. personal space

_____
_____
_____

35. primogeniture

_____
_____
_____

36. settlement landscape

_____
_____
_____

37. clustered settlements

_____
_____
_____

38. farm villages

_____
_____
_____

39. street village

_____
_____
_____

40. green village

_____
_____
_____

41. strong-point and wet-point settlements

_____
_____
_____

42. isolated farmsteads

_____
_____
_____

43. semiclustered rural settlement

_____
_____
_____

44. hamlet

_____
_____
_____

45. loose irregular and row villages

_____
_____
_____

46. Spanish-influenced architecture

_____
_____
_____

47. cenote

_____
_____
_____

48. Malthusian theory

_____
_____
_____

49. population geography

_____
_____
_____

CHAPTER 7 REVIEW: Self-evaluation Tests

PART ONE: Multiple-choice

Circle the best answer for each question. When you are finished, read each question again with your selected answer. After you are satisfied with your practice test, use the Answer Key in the back of the Study Guide to check your responses.

1. The current population of the Earth is about
   a. 2.9 billion.
   b. 9.9 billion.
   c. 4.5 billion.
   d. 12 billion.
   e. 6.2 billion.

2. Population geographers study the spatial and _____ aspects of demography.
   a. physical
   b. physiological
   c. ecological
   d. cultural
   e. distributive

3. The five most populous countries are Brazil, the United States, China, India, and
   a. Russia.
   b. Indonesia.
   c. Nigeria.
   d. Germany.
   e. Pakistan.

4. Less than 5 percent of the world's population lives in the United States.
   a. True
   b. False

5. "Moderately settled areas" have about _____ to _____ persons per square mile.
   a. 500–700
   b. 250–500
   c. 60–250
   d. 2–60
   e. None of the above

6. The density beyond which people cease to be nutritionally self-sufficient is called
   a. anatomical.
   b. physiological.
   c. nutritional.
   d. biological.
   e. caloric.

7. Birth and death rates are measured in number per
   a. hundred.
   b. thousand.
   c. hundred-thousand.
   d. million.
   e. square mile.

8. The three main population clusters of the world are eastern Asia, the Indian subcontinent, and
   a. Europe.
   b. China.
   c. the eastern United States
   d. West Africa.
   e. East Africa.

9. At present rates, how many years will it take for the world population to double?
   a. 15
   b. 20
   c. 30
   d. 40
   e. 50

10. The American tropics, North Africa, the Middle East, and Central Asia all have _____ death rates.
    a. very high
    b. high
    c. low
    d. average
    e. very low

11. Rapid population growth among humans began around the year
    a. 1450.
    b. 1930.
    c. 1890.
    d. 1820.
    e. 1700.

12. Malthus stated that famine and _____ would be constant checks of population growth.
    a. natural disasters
    b. poverty
    c. disease
    d. war
    e. accidents

13. In preindustrial societies, birth and death rates are both normally
    a. high.
    b. low.
    c. moderate.
    d. very low.
    e. None of the above.

14. Recently settled areas typically have more males than females.
    a. True
    b. False

15. Living in a farm village places the family next to their land.
    a. True
    b. False

16. The most common type of migration is
    a. forced.
    b. push-and-pull.
    c. international.
    d. war refugee.
    e. voluntary.

17. The largest migration in history took place in the _____ century when more than 50 million people left Europe.
    a. twentieth
    b. nineteenth
    c. eighteenth
    d. seventeenth
    e. sixteenth

18. In Africa, AIDS is a disease primarily affecting the _____ community.
    a. elderly
    b. upper economic
    c. heterosexual
    d. homosexual
    e. Islamic

19. Areas of extremely high heat or cold are considered to be _____ climates from the perspective of humans.
    a. "golden mean"
    b. wonderful
    c. defective
    d. marginal
    e. uninhabitable

20. Americans account for about _____ percent of the world's resources consumed each year.
    a. 5
    b. 10
    c. 15
    d. 20
    e. 40

21. The inheritance of all land by the firstborn son is termed
    a. primogeniture.
    b. Napoleonic Law.
    c. Roman Law.
    d. patrilineal.
    e. matrilineal.

22. Farm villages are common throughout much of the rural American Midwest.
    a. True
    b. False

23. The "street village" is particularly common to
    a. China.
    b. Russia.
    c. Latin America.
    d. East Africa.
    e. Southeast Asia.

24. Before the Spanish conquest, most Mayans lived in _____ villages.
    a. highland
    b. coastal
    c. isolated
    d. wet-point
    e. common

25. A "cenote" is a:
    a. village public hall.
    b. public cooking area.
    c. skate wheel.
    d. sinkhole.
    e. village elder.

PART TWO: Short answer (probable essay-type questions)

26. What do population geographers study *and* what topics and processes are included in these studies?

    _____

    _____

    _____

27. Explain the difference between population distribution and population density.

_____

_____

_____

28. Define the terms physiological density and absolute density.

_____

_____

_____

29. How is the fertility rate measured and what can this statistic indicate in regard to population patterns?

_____

_____

_____

30. What is meant by "population explosion" and how and why did this event come about?

_____

_____

_____

31. What are the contributions of Thomas Malthus to demographic studies? Are his ideas valid today?

_____

_____

_____

32. Briefly define and categorize the demographic transformation.

_____

_____

_____

33. What is a population pyramid and what are the important contributions they provide to demographers?

_____

_____

_____

34. Describe several important push-and-pull factors of migration.

_____

_____

_____

35. Provide several examples of forced migration.

_____

_____

_____

36. Briefly describe the diffusion theory of AIDS.

_____

_____

_____

37. What is population ecology?

_____

_____

_____

38. Give some examples of environmental influence on population.

_____

_____

_____

39. How does human perception play a role in distribution and settlement patterns?

_____

_____

_____

40. Label and identify the various types of farm villages. How do isolated farmsteads differ?

_____

_____

_____

CHAPTER 8

AGRICULTURAL GEOGRAPHY: FOOD FROM THE GOOD EARTH

Extended Chapter Outline (including Key Terms)

I. Agricultural regions
   A. Swidden cultivation
   B. Paddy rice farming
   C. Peasant grain, root, and livestock farming
   D. Plantation agriculture
   E. Market gardening
   F. Livestock fattening
   G. Grain farming
   H. Dairying
   I. Nomadic herding
   J. Livestock ranching
   L. Urban and nonagricultural areas

II. Agricultural diffusion
   A. Origin and diffusion of plant domestication
   B. Origin and diffusion of animal domestication
   C. Green revolution

III. Agricultural ecology
   A. Cultural adaptation
   B. Agriculturalists as modifiers of the environment
   C. Desertification
   D. Environmental perception

IV. Cultural interaction in agriculture
   A. Intensity of land use
   B. Globalization
   C. GM crops

V. Agricultural landscapes
   A. Survey, cadastral, and field patterns
   B. Fencing
   C. Hedging

VI. Conclusion and suggested readings

LEARNING OBJECTIVES OF CHAPTER 8

After reading this chapter *and* studying the maps and illustrations, you should be able to:

1. Describe various forms of agriculture such as subsistence agriculture and commercial livestock fattening.

2. Trace the origin and places of plant and animal domestication.

3. Discuss the contributions of several important scholars and geographers, most notably Sauer, Malthus, and von Thünen.

4. Explain and understand the diffusion of agriculture from one region or continent to another.

5. Describe the influence of the environment upon agricultural practices and how humans modify the earth to suit their agricultural needs and methods.

6. Understand and explain the role of human and environmental perception upon land use and agricultural production.

7. Discuss the effects of globalization and GM crops.

8. Provide examples from around the world of various agricultural lifestyles and practices.

9. Understand the various techniques of land survey, land division, fencing, and hedging that shapes the cultural landscape.

10. Continue to sharpen your skills in map reading and interpretation.

SELECTED MAP READING AND INTERPRETATION

This section of the Study Guide is intended to heighten your map reading and interpretation skills. It will also help you apply the text readings to visual and spatial display of concepts, themes, and examples in cultural geography.

A world Atlas will be very useful in completing this section of the study guide and will enhance your comprehension of the maps in the textbook. Ask your instructor to recommend an appropriate atlas to purchase (or visit the map collection at your library). A world atlas is essential for your personal reference library, not only during this course, but throughout your college career.

After reading the text and then studying the accompanying map and its captions, answer the following questions.

FIGURE 8.1 Agricultural regions of the world

1. What is the agricultural region stretching from France to Siberia, and in what other areas of the world is it found?

   _____

   _____

   _____

2. Identify several regions and countries that are dominated by livestock fattening.

   _____

   _____

   _____

3. In which regions of the world is plantation agriculture practiced? Can you think of any reason why?

   _____

   _____

   _____

4. What agricultural regions are shared by Peru, Burkina Faso, Iraq, and North Korea?

   _____

   _____

FIGURE 8.12 The origin and diffusion of agriculture

1. What are the reasons why this map should be regarded as theoretical or speculative?

   _____

   _____

   _____

2. What is the historic name of the region acknowledged as the hearth area of the earliest plant and animal domestication? What are the present day countries of this Middle Eastern region?

_____

_____

_____

3. Where is the region of domestication of vegetable reproducing plants in the Americas? What vegetables were domesticated here and what was their possible path of diffusion?

_____

_____

_____

FIGURE 8.18 Desertification in Africa

1. What countries are facing the most severe desertification?

_____

_____

_____

2. Identify the general climate types, wind patterns, and topography in the three divisions of desertification on this map.

_____

_____

_____

3. What contrasting rainfall regions exist in the yellow areas of this map? Why is desertification not a problem in many areas of the Sahara as well as the Congo Basin?

_____

_____

_____

FIGURE 8.20 Ideal and actual distribution of types of agriculture in Uruguay

1. In what ways does the spatial pattern of Uruguayan agriculture conform to von Thünen's model?

_____

_____

_____

2. How is the pattern of agriculture different from von Thünen's model, and what might cause the anomalies?

_____

_____

_____

CREATE YOUR PERSONAL GLOSSARY OF KEY TERMS, PEOPLE, AND PLACES

In the space below, write a definition and provide an example of each key term that is sufficient for **your understanding.** It is an excellent study habit to organize your response in three parts:

1. A formal definition or identification from the textbook

2. A definition of the key term in **your own words**

3. An example to add greater understanding of the key term

KEY TERMS, PEOPLE, AND PLACES FROM CHAPTER 8:

1. agriculture

   _____
   _____
   _____

2. agricultural regions

   _____
   _____
   _____

3. swidden cultivation

   _____
   _____
   _____

4. slash-and-burn agriculture

   _____
   _____
   _____

5. intertillage

   _____
   _____
   _____

6. subsistence agriculture

   _____
   _____
   _____

7. paddy rice farming

   _____
   _____
   _____

8. double-cropping

9. green revolution

10. Mediterranean agriculture

11. nomadic herding

12. plantation agriculture

13. luxury crops

14. neo-plantation

15. market gardening

16. Corn Belt

17. feedlots and commercial livestock fattening

18. commercial grain farming

_____
_____
_____

19. agribusiness

_____
_____
_____

20. commercial dairying and livestock ranching

_____
_____
_____

21. hunting and gathering groups

_____
_____
_____

22. "buffalo commons"

_____
_____
_____

23. environmental perception by agriculturalists

_____
_____
_____

24. dietary preferences

_____
_____
_____

25. agricultural productivity per capita

_____
_____
_____

26. Johann Heinrich von Thünen

_____
_____
_____

27. isolated-state model

_____
_____
_____

28. concentric zone model

_____
_____
_____

29. desertification

_____
_____
_____

30. Malthus

_____
_____
_____

31. cadastral pattern

_____
_____
_____

32. survey patterns

_____
_____
_____

33. fragmented landholding

_____
_____
_____

34. American rectangular survey system

_____
_____
_____

35. metes and bounds surveying

_____
_____
_____

36. Karl Zimmerer

_____
_____
_____

37. open-field areas

_____
_____
_____

38. hedgerow country

_____
_____
_____

39. domesticated plant

_____
_____
_____

40. Carl Sauer

_____
_____
_____

41. Fertile Crescent

_____
_____
_____

42. Mesoamerican crop complex

_____
_____
_____

43. domesticated animal

_____
_____
_____

44. "gene banks"

_____
_____
_____

45. agricultural adaptive strategies

_____
_____
_____

46. cultural adaption (agriculturalist)

_____
_____
_____

47. modification of the environment by agriculturalists

_____
_____
_____

48. GM crops

_____
_____
_____

49. The Dust Bowl

_____
_____
_____

50. cultural diffusion of the potato in Germany

_____
_____
_____

51. cattle among the Dasanetch and cultural integration

_____
_____
_____

52. threshing

_____
_____
_____

53. stone or rock fence

_____
_____
_____

CHAPTER 8 REVIEW: Self-evaluation Tests

PART ONE: Multiple-choice

Circle the best answer for each question. When you are finished, read each question again with your selected answer. After you are satisfied with your practice test, use the Answer Key in the back of the Study Guide to check your responses.

1. About _____ percent of the world's working population is employed in agriculture.
   a. 2   b. 5   c. 15   d. 30   e. 45

2. In the United States the proportion of people involved in farming is less than _____ percent.
   a. 2   b. 5   c. 10   d. 12   e. 15

3. The agricultural practice of shifting cultivation is essentially
   a. "slash-and-burn" agriculture
   b. land-rotation system.
   c. found primarily in Russia and other former communist states.
   d. subsistence agriculture.
   e. None of the above

4. The hallmark of paddy rice farming is
   a. Irish stone fences.
   b. environmental deterioration.
   c. high protein productivity.
   d. terraced paddy fields.
   e. the water buffalo.

5. Planting and harvesting the same land two or three times a year is known as
   a. intertillage.
   b. crop-rotation.
   c. double-cropping.
   d. biannual agriculture.
   e. None of the above

6. The "green revolution" involved the introduction of
   a. sugar cane.
   b. chemical fertilizers.
   c. organic methods.
   d. traditional practices of sustainability.
   e. gene banks.

7. Llamas and alpacas serve as traditional beasts of burden in
   a. South America.
   b. Mexico.
   c. Middle East and North Africa.
   d. Southeast Asia.
   e. Southeast Ohio.

8. Traditional Mediterranean agriculture is based primarily on the cultivation of barley and
   a. oats.
   b. millet.
   c. rye.
   d. wheat.
   e. rice.

9. Nomadic herding is normally practiced in what type of environment?
   a. Fertile valley floor
   b. Tropical rain forest
   c. Mountains or desert
   d. Coastal plains
   e. Temperate forests

10. Plantation agriculture usually maximizes the production of _____ for Europeans and Americans.
   a. coffee and tea
   b. grains
   c. cattle and other livestock
   d. luxury crops
   e. fruits and vegetables

11. Crops that tend to grow best in tropical highland regions are
   a. coffee and tea.
   b. grains.
   c. luxury crops.
   d. fruits and vegetables.
   e. None of the above

12. Market gardening is also referred to as
   a. neo-plantation.
   b. greenhouse farming.
   c. truck farming.
   d. market plantation.
   e. organic farming.

13. One of the most highly developed areas of commercial livestock fattening is
   a. Alpine Europe.
   b. the Mediterranean region.
   c. the Northeastern United States.
   d. the Corn Belt.
   e. Mesoamerica.

14. Fattening livestock, especially cattle, is an efficient method of protein production.
   a. True
   b. False

15. The major wheat producing countries (producing 35 percent of the world's wheat) are the United States, Russia, Canada, Argentina, Kazakhstan, and:
    a. Germany.
    b. Australia.
    c. India.
    d. Italy.
    e. Ukraine.

16. Dairy farms located near large urban areas usually produce
    a. butter.
    b. cheese.
    c. yogurt.
    d. processed milk.
    e. None of the above

17. Australia, New Zealand, South Africa, and Argentina produce 70 percent of the world's export
    a. beef.
    b. wool.
    c. coffee and tea.
    d. pork.
    e. fruit crops.

18. Geographer Carl Sauer believed that plant domestication developed as a response to hunger.
    a. True
    b. False

19. Perhaps the oldest primary region of agriculture is
    a. Southeast Asia.
    b. northwestern India.
    c. the Fertile Crescent.
    d. Mesoamerica.
    e. northeastern China.

20. The diffusion of many plant crops from the Western Hemisphere was accomplished by colonial powers, such as the Dutch and Portuguese.
    a. True
    b. False

21. Which of the following was developed to preserve what remains of domesticated plant variety?
    a. Hybrid seeds
    b. The "green revolution"
    c. Gene banks
    d. Agriculture exclusion zones
    e. Wild flower and plant nature preserves

22. A possible result of overgrazing of grasslands is
    a. deforestation.
    b. soil saturation.
    c. desertification.
    d. increased protein production.
    e. None of the above

23. The Sahel is a critical region that is located
    a. in southern India.
    b. in the Middle East.
    c. just south of the Sahara Desert.
    d. in southern Africa.
    e. in the interior of Australia.

24. Above all, most farmers rely on
    a. climatic stability.
    b. soil stability.
    c. access to water.
    d. temperature variation.
    e. low labor costs.

25. In von Thünen's isolated state model, the intensity of cultivation for any given crop increases with increasing distance from the market.
    a. True
    b. False

PART TWO: Short answer (probable essay-type questions)

26. Briefly summarize the modified version of von Thünen's isolated state model.
    _____
    _____
    _____

27. Explain the difference between the American rectangular survey system and the "metes and bounds" survey system.
    _____
    _____
    _____

28. What are the reasons for establishing "long-lot" farms and where are examples of these farms found?
    _____
    _____
    _____

29. Briefly describe the example of the cultural diffusion of the potato in Germany.

_____

_____

_____

30. What are the important themes are Carl Sauer's research, especially those concerning the domestication of plants and animals?

_____

_____

_____

31. Explain and describe the highly distinctive type of subsistence agriculture called paddy rice farming.

_____

_____

_____

32. State some of the problems as well as benefits associated with the "green revolution."

_____

_____

_____

33. Briefly describe the problems of commercial livestock fattening in regards to nutritional efficiency.

_____

_____

_____

34. What is the difference between a "suitcase farm" and agribusiness?

_____

_____

_____

35. What are the major world regions of commercial dairying?

_____

_____

_____

36. Describe the conditions and setting in which agriculture first arose.

_____

_____

_____

37. What are some of the major crops domesticated by the native peoples of the Americas?

_____

_____

_____

38. Briefly discuss the theme of cultural adaptation in agricultural geography.

_____
_____
_____
_____

CHAPTER 9

INDUSTRIES: A FAUSTIAN BARGAIN

Extended Chapter Outline (including Key Terms)

I.   Industrial regions
    A. Primary industry
    B. Secondary industry
    C. Service industry
    D. Information age

II. Diffusion of the industrial revolution
    A. Origins of the industrial revolution
    B. Diffusion from Britain

III. Industrial ecology
    A. Renewable resource crisis
    B. Acid rain
    C. The greenhouse effect and ozone depletion
    D. Radioactive pollution
    E. Response to crisis

IV. Industrial cultural interaction
    A. Labor supply
    B. Markets
    C. The political element
    D. Industrialization and cultural change

V.   Industrial landscapes

VI. Conclusion

LEARNING OBJECTIVES OF CHAPTER 9

After reading this chapter *and* studying the maps and illustrations, you should be able to:

1. Explain and define different categories of industry, including primary and secondary.

2. Trace the locations, origin, and diffusion of the industrial revolution.

3. Understand the strong effects of industrialization upon the earth's environment and its people.

4. Explain the difference between ozone depletion and the greenhouse effect, as well as the consequences of each.

5. Discuss the causal factors in selecting and determining industrial location.

6. Identify elements of various industrial landscapes and their origin.

7. Contemplate the massive impact (evident today) of the industrial revolution on the Earth, its people, and their cultures.

SELECTED MAP READING AND INTERPRETATION

This section of the Study Guide is intended to heighten your map reading and interpretation skills. It will also help you apply the text readings to visual and spatial display of concepts, themes, and examples in cultural geography.

A world atlas will be very useful in completing this section of the Study Guide and will enhance your comprehension of the maps in the textbook. Ask your instructor to recommend an appropriate atlas to purchase (or visit the map collection at your library). A world atlas is essential for your personal reference library, not only during this course, but throughout your college career.

After reading the text and then studying the accompanying map and its captions, answer the following questions.

FIGURE 9.2 Regions of selected primary and secondary industries

1. What major coastal regions show little or no commercial fishing activity? Can you think of any reasons for this?

   _____
   _____
   _____

2. What countries host the seven emerging major new centers of manufacturing?

   _____
   _____
   _____

3. Other than the United States and Russia, what countries have areas of lumbering and pulpwood?

   _____
   _____
   _____

4. What are the mining regions of Africa? What is being mined there?

   _____
   _____
   _____

FIGURE 9.3 Major regions of industry in Anglo-America

1. How would you delineate the major manufacturing belt in the United States and Canada?

   _____
   _____
   _____

143

2. Other than the major belt, what states share some of the minor regions? Why these locations?

_____

_____

_____

FIGURE 9.13 Distribution of Acid Rain in North America

1. Which states and provinces suffer the worst deposition problems? Why?

_____

_____

_____

2. Why do you think there are problems of acid precipitation in the Pacific Northwest?

_____

_____

_____

3. What is the ecology of these patterns? Include climate, weather, and topography in your answer.

_____

_____

_____

CREATE YOUR PERSONAL GLOSSARY OF KEY TERMS, PEOPLE, AND PLACES

In the space below, write a definition and provide an example of each key term that is sufficient for **your understanding.** It is an excellent study habit to organize your response in three parts:

1. A formal definition or identification from the textbook

2. A definition of the key term in **your own words**

3. An example to add greater understanding of the key term

KEY TERMS, PEOPLE, AND PLACES FROM CHAPTER 9:

1. Alfred Weber

2. industrial revolution

3. primary industry

4. secondary industry

5. service industry

6. information industry

7. ESI

8. renewable resources

_____
_____
_____

9. nonrenewable resources

_____
_____
_____

10. American Manufacturing Belt

_____
_____
_____

11. economic core and periphery

_____
_____
_____

12. uneven development

_____
_____
_____

13. technopoles

_____
_____
_____

14. deindustrialization

_____
_____
_____

15. global corporations

_____
_____
_____

16. postindustrial phase

_____
_____
_____

17. multiplier leakage

_____
_____
_____

18. high-tech corridors

_____
_____
_____

19. silicon landscapes

_____
_____
_____

20. cottage industry

_____
_____
_____

21. guild industry

_____
_____
_____

22. fossil fuels

_____
_____
_____

23. textiles

_____
_____
_____

24. metallurgy

_____
_____
_____

25. mining

_____
_____
_____

26. industrial ecology

_____
_____
_____

27. radioactive pollution

_____
_____
_____

28. Chernobyl

_____
_____
_____

29. acid rain

_____
_____
_____

30. acid precipitation

_____
_____
_____

31. greenhouse effect

_____
_____
_____

32. particulate pollutants

_____
_____
_____

33. ozone layer

_____
_____
_____

34. ozone depletion

_____
_____
_____

35. "green" reaction

_____

_____

_____

36. "green" political parties

_____

_____

_____

37. industrial inertia

_____

_____

_____

38. energy supply

_____

_____

_____

39. industrial cultural interaction

_____

_____

_____

40. labor-intensive industries

_____

_____

_____

41. weight-gaining finished product

_____

_____

_____

42. bulk-gaining finished product

_____

_____

_____

43. import-export tariffs and quotas

_____

_____

_____

44. NAFTA

_____

_____

_____

45. industrial landscape

_____

_____

_____

46. "Black Country"

_____

_____

_____

47. topocide

_____

_____

_____

48. distance in the preindustrial age

_____

_____

_____

49. Wal-Mart in China

_____

_____

_____

CHAPTER 9 REVIEW: Self-evaluation Tests

PART ONE: Multiple-choice

Circle the best answer for each question. When you are finished, read each question again with your selected answer. After you are satisfied with your practice test, use the Answer Key in the back of the Study Guide to check your responses.

1. The processing of raw materials into a more usable form is
   a. primary industry.
   b. secondary industry.
   c. tertiary industry.
   d. quaternary industry.
   e. quinary industry.

2. Agriculture is considered to be a
   a. primary industry.
   b. secondary industry.
   c. tertiary industry.
   d. quaternary industry.
   e. quinary industry.

3. The extraction of nonrenewable resources is a
   a. primary industry.
   b. secondary industry.
   c. tertiary industry.
   d. quaternary industry.
   e. quinary industry.

4. Legal services, retailing, and advertising are all considered
   a. primary industry.
   b. secondary industry.
   c. tertiary industry.
   d. quaternary industry.
   e. quinary industry.

5. Consumer-related services such as education are considered
   a. primary industry.
   b. secondary industry.
   c. tertiary industry.
   d. quaternary industry.
   e. quinary industry.

6. The American Manufacturing Belt is located
   a. from Los Angeles to New York City.
   b. primarily in the Lower South.
   c. from Chicago to Pittsburgh.
   d. around the Great Lakes and Northeast states.
   e. in New England and the Piedmont.

7. Multinational or transnational companies are also referred to as
   a. global corporations.
   b. International enterprises.
   c. Inter-regional industries.
   d. global conglomerates.
   e. None of the above

8. In Russia and Ukraine the most important mode of industrial transport is
   a. highways.
   b. railways.
   c. airways.
   d. waterways.
   e. None of the above

9. While the cottage and guild industry systems were similar in many respects, only cottage industry depended on hand labor.
   a. True
   b. False

10. The initial breakthrough in the industrial revolution occurred in the
    a. German steel industry.
    b. British smelting processes.
    c. Dutch papermaking industry.
    d. British steel industry.
    e. British textile industry.

11. The industrial revolution diffused from its place of origin first to
    a. British colonies in North America.
    b. British colonies in Africa and India.
    c. continental Europe.
    d. France.
    e. the West Midlands of England.

12. The "Faustian bargain" refers to European dependency on
    a. coal.
    b. all fossil fuels.
    c. petroleum.
    d. nuclear power.
    e. None of the above

13. The basic cause of acid rain is the burning of
    a. wood.
    b. coal.
    c. oil.
    d. fossil fuels.
    e. high sulfur coal.

14. Over 90 lakes are "dead" from acid rain in this seemingly pristine mountain range.
    a. Adirondacks
    b. Smoky Mountains
    c. Catskills
    d. Rockies
    e. Canadian Rockies

15. Greenhouse effect is caused primarily by
    a. burning coal.
    b. burning fossil fuels.
    c. ozone depletion.
    d. the hole in the ozone layer.
    e. None of the above

16. Carbon dioxide gases permit solar shortwave radiation to reach the Earth's surface and allows outgoing long-wave radiation to escape.
    a. True
    b. False

17. Ozone layer depletion is caused by
    a. fossil fuels.
    b. greenhouse effect.
    c. global warming.
    d. manufactured chemicals.
    e. radioactive waste.

18. Industrial location theory was pioneered by
    a. Carl Sauer.
    b. Max Planck.
    c. Alfred Weber.
    d. Homer Hoyt.
    e. Walter Christaller.

19. In nearly all industrial site locations, a major factor is
    a. water supply.
    b. politics.
    c. labor supply.
    d. electricity supply.
    e. climate.

20. Coastal, often scenic, fishing villages from Norway to Portugal are part of the industrial landscape.
    a. True
    b. False

PART TWO: Short answer (probable essay-type questions)

21. What and where is the origin of the industrial revolution?

_____
_____
_____

22. What was the spatial-temporal pattern of industrial diffusion?

_____
_____
_____

23. What are primary industrial activities?

_____
_____
_____

24. What is the difference between service and information industrial activities?

_____
_____
_____

25. Where is most secondary industrial activity located in North America?

_____
_____
_____

26. What was the diffusion pattern of the industrial revolution in Europe?

_____
_____
_____

27. Why did the highway displace the railroad in America as a major factor in regards to industry?

_____
_____
_____

28. What are the sources of radioactive pollution?

_____
_____
_____

29. How is acid rain created and what regions does it strongly affect?

_____
_____
_____

30. What are the major problems associated with dependency on fossil fuels?

_____

_____

_____

31. What are the main differences between greenhouse effect, ozone depletion, and global warming?

_____

_____

_____

32. Explain the causal factors of greenhouse effect.

_____

_____

_____

33. What is meant by the "green" reaction?

_____

_____

_____

34. What are the pioneering contributions to industrial location theory advanced by Alfred Weber?

_____

_____

_____

35. Describe some environmental factors in industrial location.

_____

_____

_____

36. Describe the various relationships between labor and industrial location.

_____

_____

_____

37. What all encompassing cultural changes were brought about by the industrial revolution?

_____

_____

_____

38. Explain the various impacts politics can have on industry.

_____

_____

_____

39. Describe industrial landscape from the categories of primary, secondary, and quaternary industries.

_____

_____

_____

40. Provide examples of how the industrial landscape is interpreted by some humanists.

_____

_____

_____

CHAPTER 10

URBANIZATION: THE CITY IN TIME AND SPACE

Extended Chapter Outline (including Key Terms)

I. Culture region
    A. Patterns and processes
    B. Impacts

II. Origin and diffusion of the city
    A. Single-factor and multiple-factor models for the rise of cities
    B. Urban hearth areas
    C. Diffusion of the city from hearth areas

III. Evolution of urban landscapes: Global cities
    A. The Greek city
    B. Roman cities
    C. The medieval city
    D. The Renaissance and Baroque periods
    E. The capitalist city
    F. The colonial city
    G. Class, "race," and gender in the industrial city
    H. Megalopolis
    I. Edge cities
    J. Globalizing cities in the developing world

IV. The ecology of urban location
    A. Site and situation
    B. Defensive sites
    C. Trade-route sites

V. Cultural interaction in urban geography

VI. Conclusion

LEARNING OBJECTIVES OF CHAPTER 10

After reading this chapter *and* studying the maps and illustrations, you should be able to:

1. Define the "city" and discuss the meaning of urbanized population.

2. Explain various theories dealing with the origin and diffusion of the city.

3. Demonstrate a knowledge of urban hearth areas and the ten largest urban centers today.

4. Discuss the evolution of urban landscapes.

5. Understand the relationship between functional zonation and urban morphology.

6. Explain the causal factors that led to the capitalist city.

7. Discuss the impact of industrialization on city structure.

8. Understand various models of cities in developing countries.

9. Delineate the types of city sites and their functions.

10. Interpret with new understanding various cities with which you are familiar.

CREATE YOUR PERSONAL GLOSSARY OF KEY TERMS, PEOPLE, AND PLACES

In the space below, write a definition and provide an example of each key term that is sufficient for **your understanding.** It is an excellent study habit to organize your response in three parts:

1. A formal definition or identification from the textbook

2. A definition of the key term in **your own words**

3. An example to add greater understanding of the key term

KEY TERMS, PEOPLE, AND PLACES FROM CHAPTER 10:

1. urbanized population

_____
_____
_____

2. rural-to-city migration

_____
_____
_____

3. World's 20 largest metropolitan areas

_____
_____
_____

4. primate city

_____
_____
_____

5. single-factor model

_____
_____
_____

6. hydraulic civilization

_____
_____
_____

7. institution of kingship

_____
_____
_____

8. multiple-factor models

_____

_____

_____

9. urban hearth areas

_____

_____

_____

10. axis mundi

_____

_____

_____

11. walled inner city

_____

_____

_____

12. urban morphology

_____

_____

_____

13. functional zonation

_____

_____

_____

14. acropolis and agora

_____

_____

_____

15. Etruscans

_____

_____

_____

16. forum

_____

_____

_____

17. Byzantine civilization

_____

_____

_____

18. the medieval period
_____
_____
_____

19. city "charter"
_____
_____
_____

20. suburbs
_____
_____
_____

21. cathedral
_____
_____
_____

22. guildhall
_____
_____
_____

23. Renaissance period
_____
_____
_____

24. Baroque period
_____
_____
_____

25. the capitalist city
_____
_____
_____

26. industrial city
_____
_____
_____

27. laissez-faire utilitarianism
_____
_____
_____

28. benefits of agglomeration

_____
_____
_____

29. zoning

_____
_____
_____

30. "class" and "race" and "gender"

_____
_____
_____

31. megalopolis

_____
_____
_____

32. "global cities"

_____
_____
_____

33. Islamic world cities

_____
_____
_____

34. the colonial city

_____
_____
_____

35. gridiron street pattern

_____
_____
_____

36. barriadas

_____
_____
_____

37. squatter settlements

_____
_____
_____

38. "site" and "situation"

_____

_____

_____

39. river-island site

_____

_____

_____

40. acropolis sites

_____

_____

_____

41. trade-route sites

_____

_____

_____

42. spatial distribution of cities

_____

_____

_____

43. "central-place" theory

_____

_____

_____

44. tertiary stage of production

_____

_____

_____

45. Walter Christaller

_____

_____

_____

46. "threshold" and "range"

_____

_____

_____

47. hinterlands

_____

_____

_____

48. fifteenth-century humanism

_____
_____
_____

49. Calcutta

_____
_____
_____

CHAPTER 10 REVIEW: Self-evaluation Tests

PART ONE: Multiple-choice

Circle the best answer for each question. When you are finished, read each question again with your selected answer. After you are satisfied with your practice test, use the Answer Key in the back of the Study Guide to check your responses.

1. Urbanized population refers to a country's
    a. number of cities.
    b. number of cities over 100,000 in population.
    c. percentage of urban population.
    d. ratio of urban to rural population.
    e. None of the above

2. The United States Census Bureau defines a city as a densely populated area of _____ people or more.
    a. 2,500
    b. 5,000
    c. 10,000
    d. 20,000
    e. 50,000

3. The term "world cities" refers to cities having populations of at least
    a. 500,000.
    b. 1 million.
    c. 5 million.
    d. 10 million.
    e. None of the above

4. Which of the following cities is *not* ranked in the world's ten largest metropolitan areas?
    a. Seoul
    b. London
    c. Moscow
    d. New York
    e. Bombay

5. The city that dominates the political, economic, and cultural life of a country is the _____ city.
    a. capital
    b. most populous
    c. vanguard
    d. primate
    e. growth pole

6. The origin of cities is strongly related to
   a. plant and animal domestication.
   b. improved transportation networks.
   c. warfare and defense.
   d. improved building techniques.
   e. All of the above

7. The hydraulic civilization model can be tied to all of the following urban hearths *except*
   a. China.
   b. Mesoamerica.
   c. Egypt.
   d. Mesopotamia (Iraq).
   e. the Indus River valley.

8. Cosmomagical cities exhibit the three spatial characteristics of a _____,
   symbolic center, and cardinal direction orientation.
   a. walled circumference
   b. walled outer city
   c. universe-like form
   d. wheel spoke form
   e. moat or bulwark

9. The pattern of functional land use within a city is referred to as
   a. urban morphology.
   b. functional development.
   c. functional zonation.
   d. spatial landscapes.
   e. urban land use .

10. The place for public use and markets in ancient Greek cities was the
    a. agora.
    b. citadel.
    c. forum.
    d. acropolis.
    e. castra.

11. The _____ was the urban zone of a Roman city where religious, administrative, and
    educational structures, as well as markets, were located.
    a. agora
    b. citadel
    c. forum
    d. acropolis
    e. castra

12. The most important contribution or legacy of Roman city builders was probably
    a. engineering.
    b. transportation.
    c. architecture.
    d. site selection.
    e. gridiron planning.

13. The major functions of the medieval city are depicted in all of the following symbols *except* the
    a. town hall.
    b. charter.
    c. fortress.
    d. wall.
    e. marketplace.

14. The medieval town's crowning glory was usually the
    a. town hall.
    b. castle.
    c. fortress.
    d. cathedral.
    e. clock tower.

15. Most European cities were founded during the medieval period.
    a. True
    b. False

16. Few of the traditions of Western urban life began in the medieval period.
    a. True
    b. False

17. The medieval landscape is still with us, giving us a visible history of the city and distinctive form.
    a. True
    b. False

18. The Baroque period predated the Renaissance in history.
    a. True
    b. False

19. The term and concept of megalopolis was created by
    a. Walter Christaller.
    b. Walter Kollmorgen.
    c. Lewis Mumford.
    d. Alfred Weber.
    e. Jean Gottmann.

20. A majority of the American people were not urbanized until
    a. 1890.
    b. 1920.
    c. 1940.
    d. 1950.
    e. 1960.

21. The regional setting of an urban location is called the
    a. site.
    b. node.
    c. ecological niche.
    d. situation.
    e. None of the above

22. Mexico City, Montreal, New York City, and Venice are all examples of
    a. limited-access cities.
    b. colonial cities.
    c. river-island sites.
    d. acropolis sites.
    e. defensive sites.

23. Critical to central-place theory is the fact that different goods and services vary both in
    _____ and range.
    a. spatial distribution
    b. quality
    c. quantity
    d. threshold
    e. access

24. An "edge city" is any place that has the following features, *except*
    a. 1 million square feet or less of leasable office space.
    b. 600,000 square feet or more of leasable retail space.
    c. more jobs than bedrooms.
    d. perceived by people as one place.
    e. was not at all urban as recently as 30 years ago.

25. Large scale squatter settlements are a typical feature of many
    a. Mexican cities.
    b. European and Russian cities.
    c. "developing" world cities.
    d. former French and Portuguese colonies.
    e. North African and Middle Eastern cities.

PART TWO: Short answer (probable essay-type questions)

26. Briefly explain "threshold" and "range" in central-place theory.

_____

_____

_____

27. What did Christaller add to central-place theory in his second model?

_____

_____

_____

28. List some common features of former colonial cities.

_____

_____

_____

29. What are some of the major problems faced by cities in Africa, Mesoamerica, and India?

_____

_____

_____

30. Describe a squatter settlement, including a couple of actual examples.

_____

_____

_____

31. Clearly explain the difference between city site and situation.

_____

_____

_____

32. Describe the alteration of Paris during the nineteenth century.

_____

_____

_____

33. What is meant by "capitalist city"?

_____

_____

_____

34. What role does "race" play in urban function and features?

_____

_____

_____

35. Explain "laissez-faire utilitarianism" and its effects on cities.

_____
_____
_____

36. Describe some main features of a medieval town.

_____
_____
_____

37. What are the spatial features associated with cosmomagical cities?

_____
_____
_____

38. Discuss an argument concerning the diffusion of the city from hearth areas.

_____
_____
_____

39. What are the concepts of urban morphology and functional zonation?

_____
_____
_____

40. What was a medieval town "charter" and what were its implications?

_____
_____
_____

CHAPTER 11

INSIDE THE CITY: A CULTURAL MOSAIC

Extended Chapter Outline (including Key Terms)

I. Urban culture regions
    A. Social regions
    B. Neighborhoods

II. Cultural diffusion in the city
    A. Centralization
        1. Economic and social advantages
    B. Suburbanization and decentralization
        1. Economic and social advantages
        2. Public policy
    C. The costs of decentralization
    D. Gentrification

III. Cultural ecology of the city
    A. Urban weather and climate
    B. Urban hydrology
    C. Urban vegetation

IV. Cultural interaction and models of the city
    A. Concentric zone model
    B. Sector model
    C. Multiple nuclei model
    D. Feminist critiques
    E. Apartheid and postapartheid cities
    F. Soviet and post-Soviet cities
    G. Latin American model

V. Urban landscapes
    A. Themes in cityscape study
        1. Landscape dynamics
        2. The city as palimpsest
        3. Symbolic cityscapes
        4. Perception of the city
    B. The new urban landscape

VI. Conclusion

LEARNING OBJECTIVES OF CHAPTER 11

After reading this chapter *and* studying the maps and illustrations, you should be able to:

1. Identify and describe various and distinct urban culture regions.

2. Understand the concept, views, perception, and importance of the "neighborhood" and its role in everyday life.

3. Explain the difference between and the pros and cons of centralization and decentralization of the city.

4. Understand the elements of the urban ecosystem and its impact on society.

5. Identify various city models and relate them to actual cities.

6. Understand the elements of the city landscape and its causal factors.

7. Identify symbolic, cultural, and perceptive elements of the city environment.

8. Begin to look and interpret, with new awareness, the cities and neighborhoods around you.

CREATE YOUR PERSONAL GLOSSARY OF KEY TERMS, PEOPLE, AND PLACES

In the space below, write a definition and provide an example of each key term that is sufficient for **your understanding.** It is an excellent study habit to organize your response in three parts:

1. A formal definition or identification from the textbook

2. A definition of the key term in **your own words**

3. An example to add greater understanding of the key term

KEY TERMS, PEOPLE, AND PLACES FROM CHAPTER 11:

1. social culture region

   _____
   _____
   _____

2. ethnic culture region

   _____
   _____
   _____

3. census tracts

   _____
   _____
   _____

4. "neighborhood"

   _____
   _____
   _____

5. definitions of homelessness

   _____
   _____
   _____

6. "zone in transition"

   _____
   _____
   _____

7. centralizing forces

   _____
   _____
   _____

8. decentralizing forces

9. economic and social advantages of centralization

10. agglomeration or clustering

11. "streetcar suburbs"

12. decentralization

13. economic and social advantages of decentralization

14. lateral commuting

15. Federal Housing Administration

16. covenants

17. red-lining

18. costs of decentralization

_____
_____
_____

19. gentrification

_____
_____
_____

20. sexuality and gentrification

_____
_____
_____

21. costs of gentrification

_____
_____
_____

22. urban ecosystem

_____
_____
_____

23. urban geology

_____
_____
_____

24. urban heat island

_____
_____
_____

25. urban hydrology

_____
_____
_____

26. runoff

_____
_____
_____

27. urban vegetation

_____
_____
_____

28. segregation

_____
_____
_____

29. social group "invasion" and "succession"

_____
_____
_____

30. concentric zone model

_____
_____
_____

31. CBD

_____
_____
_____

32. sector model

_____
_____
_____

33. Homer Hoyt

_____
_____
_____

34. Ernest Burgess

_____
_____
_____

35. multiple-nuclei models

_____
_____
_____

36. feminist critique

_____
_____
_____

37. Latin American model

_____
_____
_____

38. spine/sector

_____
_____
_____

39. Soviet and post-Soviet cities

_____
_____
_____

40. zone of accretion

_____
_____
_____

41. zone of peripheral squatter settlements

_____
_____
_____

42. urban landscape and cityscape

_____
_____
_____

43. landscape dynamics

_____
_____
_____

44. city as palimpsest

_____
_____
_____

45. symbolic cityscape

_____
_____
_____

46. ornamental iconography

_____
_____
_____

47. perception of the city

_____
_____
_____

48. landmarks

49. office parks

50. high-tech corridors

51. festival settings

52. militarized space

53. public space

CHAPTER 11 REVIEW: Self-evaluation Tests

PART ONE: Multiple-choice

Circle the best answer for each question. When you are finished, read each question again with your selected answer. After you are satisfied with your practice test, use the Answer Key in the back of the Study Guide to check your responses.

1. An excellent method of defining social regions is to isolate one social trait and plot its distribution within a city by using
   a. aerial maps
   b. satellite imagery.
   c. census tracts.
   d. telephone directories.
   e. property tax records.

2. The neighborhood concept is critical to cultural geography because it
   a. recognizes the sentiment people have for places.
   b. provides examples of race relations within cities.
   c. illustrates and explores urban ecology.
   d. helps us understand spatial structures of cities.
   e. None of the above

3. An important economic advantage to central city location has traditionally been
   a. product cost.
   b. advertising.
   c. close to major transport networks.
   d. accessibility.
   e. All of the above

4. Agglomeration or clustering is
   a. considered an economic disadvantage.
   b. considered an economic advantage.
   c. considered the hope for the future of neighborhoods.
   d. considered an out-of-date method of urban planning.
   e. only used in North American cities.

5. Today, many people travel to work from suburb to suburb. This is termed
   a. suburban dependency.
   b. "by-pass commuting."
   c. lateral commuting.
   d. highway by-pass commuting.
   e. None of the above

6. The government body known as the "FHA" is an abbreviation for
   a. Fair Housing Act.
   b. Federal Home Association.
   c. Family Housing Act.
   d. Federal Housing Act.
   e. Federal Housing Administration.

7. A practice in which banks and mortgage companies designate areas considered to be a high risk for loans is
   a. redistricting.
   b. red-lining.
   c. covenants.
   d. black listing.
   e. None of the above

8. The movement of middle-class people into deteriorated areas of a city center is referred to as
   a. urban revitalization.
   b. urban renewal.
   c. gentrification.
   d. insanity.
   e. reverse out-migration.

9. Lower-income people of the inner city are often displaced by the process of
   a. urban sprawl.
   b. urban housing projects.
   c. gentrification.
   d. homelessness.
   e. reverse out-migration.

10. The study of the relationship between an organism and its physical environment is called
   a. physical geography.
   b. physical geology.
   c. Earth Science.
   d. ecology.
   e. cultural ecology.

11. The direction of city growth, the patterning of social regions, and the routing of transportation can all be influenced by
   a. topography.
   b. river and waterway patterns.
   c. climate.
   d. seasonality.
   e. temperature variation.

12. The heat generation of a city produces a large mass of warmer air sitting over the city called the urban heat island.
   a. True
   b. False

13. During the summer, a city center is warmer than its suburbs.
    a. True
    b. False

14. The concentric zone model of Burgess has five zones. Zone 2, characterized by a mixed pattern of industrial and residential land use, is considered a _____ zone.
    a. stable
    b. blue-collar
    c. transitional
    d. better housing
    e. gentrified

15. The area of the concentric zone model characterized by commuters and high-income families is zone _____ .
    a. 1
    b. 2
    c. 3
    d. 4
    e. 5

16. The sector model of urban land-use was developed by
    a. Burgess.
    b. Hoyt.
    c. Harris.
    d. Ullman.
    e. None of the above

17. Freeways became part of the city landscape
    a. after 1945.
    b. before 1940.
    c. since the 1950s.
    d. only since the 1960s.
    e. before 1930.

18. The model that maintains that a city develops with equal intensity around various points is the _____ model.
    a. sector
    b. concentric zone
    c. urban growth
    d. multiple nuclei
    e. None of the above

19. In the Latin American model the zone of maturity primarily includes
    a. traditional colonial homes.
    b. diverse housing types.
    c. new industrialization.
    d. squatter settlements.
    e. business and commerce.

20. A palimpsest is a
    a. section of most American cities.
    b. section of most European cities.
    c. critical part of the sector model.
    d. reference to transport planning.
    e. parchment used over and over for written messages.

21. A cemetery may be considered part of a city's
    a. park system.
    b. symbolic landscape.
    c. palimpsest.
    d. CBD.
    e. All of the above

22. Which of the following is considered a "landmark"?
    a. City hall
    b. Railway depot
    c. Industrial smokestack
    d. White Castle restaurant
    e. All of the above

23. Shopping malls in North America are _____ spaces.
    a. eternal
    b. urban
    c. private
    d. open
    e. public

24. Mike Davis discusses patterns of "militarized" space. This space is primarily
    a. military administration sections of cities.
    b. military port urban zones.
    c. any military base area, urban or rural.
    d. defensive urban space.
    e. None of the above

25. Many U.S. cities have a pattern of leap-frog expansion, and then in-filling, to use all of the space inside the city limits.
    a. True
    b. False

PART TWO: Short answer (probable essay-type questions)

26. What human factors define a neighborhood?

_____
_____
_____

27. What is meant by the "city as palimpsest"?

_____
_____
_____

28. Describe elements of a neighborhood with which you are familiar.

_____
_____
_____

29. What are the characteristics of the Latin American model zones?

_____
_____
_____

30. What are the elements of Homer Hoyt's urban model?

_____
_____
_____

31. How is the focus of this chapter different from that of the previous chapter?

_____
_____
_____

32. List some symbols and landmarks of the urban environment.

_____
_____
_____

33. How do cities affect the natural environment and ecology?

_____
_____
_____

34. What does Kevin Lynch mean by "perception of the city"?

_____
_____
_____

35. What are some criticisms of the concentric zone model?

_____

_____

_____

36. What is the "feminist critique" in urban geography?

_____

_____

_____

37. Briefly discuss the role of race and gender in the urban mosaic.

_____

_____

_____

38. Describe the concept of the shopping mall as a social center.

_____

_____

_____

39. What are some of the costs of city decentralization?

_____

_____

_____

40. How did the streetcar and trolley affect neighborhood development in the late-nineteenth and early twentieth centuries?

_____

_____

_____

41. Are neighborhoods found in the suburbs? Why or why not?

_____

_____

_____

42. What are some of the human costs of gentrification?

_____

_____

_____

43. How does topography influence city growth and structure?

_____

_____

_____

44. Is urban hydrology a factor in the future of cities? How and why?

_____
_____
_____

45. What are the problems of defining "homelessness"?

_____
_____
_____

46. Describe the composition of the new ethnic neighborhoods.

_____
_____
_____

CHAPTER 12

ONE WORLD OR MANY? THE CULTURAL GEOGRAPHY OF THE FUTURE

Extended Chapter Outline

I.   Cultural geography of the future

II. Globalization
    A. History, geography, and the globalization of everything
    B. Globalization and its discontents
    C. Many worlds

III. Culture regions
    A. Uneven geography
    B. One Europe or many?
    C. Glocalization
    D. Geography of the internet

IV. Cultural diffusion
    A. Information superhighway
    B. The internet
    C. American car culture

V. Cultural ecology
    A. Sustainability
    B. Think globally, act locally

VI. Cultural interaction
    A. Global tourist
    B. World music

VII. Cultural landscape
    A. Globalized
    B. Striving for the unique
    C. Wal-Martians invade!
    D. Europe's rural landscape

VIII. Conclusion

Lessons and Purpose of Chapter 12

This chapter is somewhat different from the previous sections of your text. It is a chapter about the geography of the future. We are all interested in the future because that is where we will spend the rest of our lives.

In lieu of practice tests and key terms, the following section of major essay questions is provided for your thoughts and notes

1. What is meant by the "End of Geography" in relation to globalization? What will be the landscapes of the future?

_____
_____
_____
_____
_____
_____
_____
_____
_____
_____
_____
_____
_____
_____
_____
_____
_____
_____
_____
_____
_____
_____
_____
_____
_____
_____
_____
_____
_____

2. Describe various critiques and actions both pro and con globalization.

_____

_____

_____

_____

_____

_____

_____

_____

_____

_____

_____

_____

_____

_____

_____

_____

_____

3. Discuss how media, human cybermobility, and the Internet challenge change geography.

_____

_____

_____

_____

_____

_____

_____

_____

_____

_____

_____

_____

_____

_____

_____

_____

_____

_____

_____

_____

_____

ANSWER KEY

CHAPTER 1
1. d	6. a	11. a	16. c	21. c
2. b	7. c	12. a	17. a	22. e
3. d	8. b	13. c	18. d	23. a
4. a	9. a	14. c	19. b	24. b
5. d	10. b	15. b	20. a	

CHAPTER 2
1. c	9. d	17. a	25. e	33. d
2. a	10. a	18. c	26. d	34. c
3. d	11. e	19. c	27. c	35. b
4. b	12. e	20. a	28. d	36. a
5. e	13. c	21. e	29. a	37. c
6. b	14. e	22. e	30. b	
7. c	15. d	23. a	31. d	
8. c	16. b	24. a	32. b	

CHAPTER 3
1. b	7. d	13. d	19. a	25. e
2. c	8. e	14. d	20. a	26. c
3. e	9. e	15. b	21. a	
4. d	10. b	16. b	22. a	
5. c	11. a	17. c	23. b	
6. a	12. d	18. a	24. d	

CHAPTER 4
1. d	6. d	11. e	16. c	21. a
2. e	7. d	12. b	17. d	22. b
3. c	8. b	13. c	18. a	
4. e	9. a	14. c	19. e	
5. a	10. a	15. d	20. d	

CHAPTER 5
1. c	5. e	9. b	13. d	17. c
2. b	6. b	10. c	14. b	18. d
3. a	7. b	11. a	15. a	19. c
4. e	8. d	12. b	16. e	20. a

CHAPTER 6
1. d	6. b	11. c	16. a	21. b
2. a	7. b	12. e	17. b	22. e
3. c	8. d	13. a	18. e	23. d
4. b	9. b	14. b	19. c	24. a
5. d	10. b	15. e	20. d	25. d

CHAPTER 7
1. e	6. b	11. e	16. e	21. a
2. c	7. b	12. d	17. b	22. b
3. b	8. a	13. a	18. c	23. b
4. a	9. e	14. a	19. d	24. d
5. c	10. c	15. b	20. d	25. d

CHAPTER 8
1. e	6. b	11. a	16. d	21. c
2. a	7. a	12. c	17. b	22. c
3. b	8. d	13. d	18. b	23. c
4. d	9. c	14. b	19. c	24. a
5. c	10. d	15. e	20. a	25. b

CHAPTER 9
1. b	5. e	9. b	13. d	17. d
2. a	6. d	10. e	14. a	18. c
3. a	7. a	11. c	15. b	19. c
4. d	8. b	12. b	16. b	20. a

CHAPTER 10
1. c	6. a	11. c	16. b	21. d
2. a	7. b	12. d	17. a	22. e
3. d	8. c	13. a	18. b	23. d
4. b	9. c	14. d	19. e	24. a
5. d	10. a	15. a	20. b	25. c

CHAPTER 11
1. c	6. e	11. a	16. b	21. b
2. a	7. b	12. a	17. a	22. e
3. d	8. c	13. a	18. d	23. c
4. b	9. c	14. c	19. a	24. d
5. c	10. d	15. e	20. e	25. a